101 FACTS
YOU DIDN'T KNOW ABOUT
DEEP SPACE

'In memory or Mum and Dad who supported my dream to reach out among the stars'

101 FACTS
YOU DIDN'T KNOW ABOUT
DEEP SPACE
Beyond our Solar System

MARK THOMPSON

WHITE OWL
AN IMPRINT OF PEN & SWORD BOOKS LTD.
YORKSHIRE – PHILADELPHIA

First published in Great Britain in 2026
White Owl
An imprint of
Pen & Sword Books Ltd
Yorkshire – Philadelphia

Print ISBN 978 1 03611 145 8
ePub ISBN 978 1 03611 147 2
ePDF ISBN 978 1 03611 149 6

A CIP catalogue record for this book is available from the British Library.

Typeset in 9.6/12.6 pts Fira Sans
by SJmagic DESIGN SERVICES, India.
Printed and bound in India by Replika Press Pvt. Ltd

The Publisher's authorised representative in the EU for product safety is Authorised Rep Compliance Ltd., Ground Floor, 71 Lower Baggot Street, Dublin D02 P593, Ireland.
www.arccompliance.com

For a complete list of Pen & Sword titles please contact

PEN & SWORD BOOKS LIMITED
George House, Beevor Street, Off Pontefract Road, Hoyle Mill, Barnsley,
South Yorkshire, England, S71 1HN.
E-mail: enquiries@pen-and-sword.co.uk
Website: www.pen-and-sword.co.uk

or

PEN AND SWORD BOOKS
1950 Lawrence Rd, Havertown, PA 19083, USA
E-mail: uspen-and-sword@casematepublishers.com
Website: www.penandswordbooks.com

MIX
Paper from
responsible sources
FSC® C016779

CONTENTS

FACT 1: The Milky Way Smells of rum and raspberries! 8

FACT 2: Exoplanets' light is studied to learn about their atmosphere..................10

FACT 3: Rogue planets wander freely through the Galaxy 12

FACT 4: Gamma-ray bursts can release more energy in 10 seconds than our Sun will in its entire life 14

FACT 5: Some galaxies are so far away that we will never be able to see them 16

FACT 6: Most galaxies in the universe are dwarf galaxies..18

FACT 7: Stars do not actually twinkle.... 20

FACT 8: The universe is expanding at an accelerated rate...................................22

FACT 9: Water is common in space 24

FACT 10: The first exoplanet was discovered in 1992..26

FACT 11: We have seen a galaxy whose light travelled 33 billion years to reach us.. 28

FACT 12: We have successfully redirected an asteroid.................................. 30

FACT 13: Early galaxies look nothing like what we see today.................................32

FACT 14: There is very little dark matter at the core of the Milky Way.......34

FACT 15: The Kuiper belt of icy objects is bigger than we thought.......................... 36

FACT 16: Betelgeuse's surface is boiling furiously 38

FACT 17: Neutron stars are jetting material away at 40% the speed of light.. 40

FACT 18: One teaspoon of neutron star material would weigh the same as 2.6 billion people 42

FACT 19: The largest supercluster is home to 26 quadrillion stars................... 44

FACT 20: Barred spiral galaxies appeared early, just 2.7 billion years after the Big Bang 46

FACT 21: Black holes can kill galaxies.... 48

FACT 22: We can measure the magnetic field of distant planets using radio telescopes 50

FACT 23: We can see the echoes from the Big Bang 52

FACT 24: Some astronomers think a signal from ET was received in 1977 54

FACT 25: The Milky Way might have too many satellite galaxies 56

FACT 26: Fish could turn regolith into fertile soil on Mars............................. 58

FACT 27: It is possible that heart attacks may be more common in space 60

FACT 28: Merging galaxies make for explosive star formation 62

FACT 29: Many planets have the raw ingredients for life 64

FACT 30: There is a star that measures 11.4 kilometres across............ 66

FACT 31: Stars' dwarf galaxies help us to map dark matter 68

FACT 32: We have a way to see whether the universe has been designed to support life....70

FACT 33: There is a technique for spotting Dyson rings....72

FACT 34: Pulsars are the ideal probes for dark matter....74

FACT 35: There is a distinctive background glow of light across the universe....76

FACT 36: Oxygen in deep alien oceans of water worlds can sustain life....78

FACT 37: A star (possibly) has been discovered and it is escaping the Milky Way....80

FACT 38: Gravitational wave observatories can detect collapsing stars....82

FACT 39: Iron winds are blowing on WASP-76b....84

FACT 40: Slime can teach us about the cosmic web....86

FACT 41: The longest black hole jet is 23 million light years....88

FACT 42: Scientists have simulated nuclear detonations in space....90

FACT 43: A black hole has been seen eating a star....92

FACT 44: Stars can get hiccups....94

FACT 45: You are made of carbon that took a journey into intergalactic space....96

FACT 46: Scientific models now support the potential stability of Dyson spheres....98

FACT 47: Runaway stars can escape the Milky Way....100

FACT 48: We are unlikely to be able to detect advanced civilisations by their industrial pollution....102

FACT 49: We have sent golden records out into deep space....104

FACT 50: Gold is created in deep space....106

FACT 51: Astronauts cannot cry properly because there's no gravity to pull tears away....108

FACT 52: Some stars have surface temperatures cooler than your morning coffee....110

FACT 53: There are more stars in the Milky Way than humans *ever* born!....112

FACT 54: Only a fraction of stars are white....113

FACT 55: There are diamonds scattered across the galaxy....114

FACT 56: A planet has been found clinging to a star racing through space at over 500 kilometres per second....116

FACT 57: Jumping on deep space trips helps stop muscles from wasting away....118

FACT 58: X-ray emissions reveal a planet has been destroyed by a white dwarf....120

FACT 59: You age slower when travelling faster....122

FACT 60: Exocomets exist around distant stars....124

FACT 61: Astronomers still do not know why more galaxies spin one way than the other....126

FACT 62: Exoplanets can be discovered by listening for radio waves....128

FACT 63: There are pumpkin-shaped stars....130

FACT 64: The universe has no centre....132

FACT 65: Black holes can evaporate.......134

FACT 66: The solar system is moving through space at 828,000 kilometres per hour.......136

FACT 67: The universe will eventually become cold, dark and empty.......138

FACT 68: There is a galaxy with no stars.......140

FACT 69: The Boomerang Nebula is the coldest place in the universe.......142

FACT 70: Black holes can sing.......144

FACT 71: The universe is getting darker.......146

FACT 72: There are natural telescopes in deep space.......148

FACT 73: We cannot see 27% of the universe.......150

FACT 74: Galaxies are colliding in space.......152

FACT 75: There is a massive void in space.......154

FACT 76: Astronauts can smell seared steak after a space walk.......156

FACT 77: Neutron stars have mountains just millimetres tall.......158

FACT 78: The universe has a beginning and will have an end.......160

FACT 79: Most stars are red dwarfs.......162

FACT 80: Galaxies have habitable zones too.......164

FACT 81: Space ripples like the ocean.......166

FACT 82: The largest black hole is as massive as 17 billion suns.......168

FACT 83: Nothing can travel faster than the speed of light, except the universe.......170

FACT 84: Our solar system is surrounded by a bubble.......172

FACT 85: Water ice has been found on an exoplanet.......174

FACT 86: Planets are forming at the centre of our Galaxy.......176

FACT 87: Black holes are firing gas out into space.......178

FACT 88: Stars can orbit inside other stars.......180

FACT 89: The Hercules-Corona Borealis Great Wall is the largest structure in the universe.......182

FACT 90: There's a quasar that outshines 600 trillion suns.......184

FACT 91: Cosmic rays can mutate human beings.......186

FACT 92: It is unusual, but some stars form in complete isolation.......188

FACT 93: There are some black holes that are older than galaxies.......190

FACT 94: Some exoplanets orbit dead stars.......192

FACT 95: Organic molecules exist in interstellar gas clouds.......194

FACT 96: The universe may be shaped like a doughnut.......196

FACT 97: There is an exoplanet where it rains molten glass.......198

FACT 98: Stars can be 'spaghettified' by black holes.......200

FACT 99: Some white dwarfs can reignite and explode.......202

FACT 100: There are supersonic winds on exoplanets.......204

FACT 101: The Milky Way has a 'fossil' galaxy inside it.......206

FACT 1

The Milky Way Smells of rum and raspberries!

Space is big, really big, and it is full of really quite amazing wonders. It is reasonable to consider that galaxies are the building blocks of the universe. They can be likened to great big islands in the vastness of space where stars, planets, moons, comets, star clusters and gas clouds are held together under the force of gravity.

I often like to muse that galaxies do not really exist because they are simply collections of other stuff. My home town of Norwich in the UK is a collection of houses, cars, shops, offices, parks, people, etc. and the list goes on. If you took all of those things away then Norwich would not actually exist and it is the same with galaxies – take away all the things that they are composed of and they just would not exist! Back to the fact.

Our own Galaxy (note the upper case 'G' in the spelling of galaxy mid-sentence denotes our own Milky Way Galaxy, rather than any other galaxy) is called the Milky Way and is about 100,000 light years across. A light year is the distance light can travel in one year at a velocity of 300,000 kilometres per second so is a measure of distance not time. One light year equates to 9.5 trillion kilometres.

Galaxies are full of dust clouds and in particular if you want to study the core then it is very difficult with optical telescopes. Instead astronomers turn to different wavelengths of 'light' to explore the darkest, deepest secrets at the centre of the Milky Way. In particular, radio wavelengths are suitable as they can pierce through the clouds.

A team of astronomers using radio telescopes has been exploring the giant molecular cloud found near the core of our Galaxy located just 390 light years from the centre. It is known as Sagittarius B2 and is one of the largest known molecular clouds in the Milky Way. The team found evidence of a chemical known as ethyl formate which occurs naturally in ants and in the stingers of bees, but of course they are not buzzing around the Galaxy. What is exciting and enchanting about the discovery is that ethyl formate is also responsible for the flavour in raspberries and smells like rum. Seems that the centre of the Galaxy is the place to be for a galactic knees-up.

Image of the Milky Way above Paranal in Chile taken by ESO astronomer ESO astronomer Yuri Beletsky. (ESO/Y. Beletsky)

FACT 2

Exoplanets' light is studied to learn about their atmosphere

We are all very familiar with this thing we call light after all – our eyes use it to explore the world around us. Light can be thought of as a wave with peaks and troughs just like a wave on the ocean. Visible light has a wavelength – the distance between successive peaks or successive troughs – of 4 to 7 millionths of a metre. The light that we see is only a tiny portion of something called the electromagnetic spectrum and the only thing that differentiates one type of radiation from another is the wavelength. At the longer wavelength end are the radio waves and, as the wavelength decreases, microwaves, infrared, visible light, ultraviolet, X-rays and finally gamma rays at the shortest wavelength.

Objects in the universe generally emit radiation in all forms and to fully understand the nature of the object we study all the radiation. It is a little like listening to music. If you just listen to the strings, you do not fully appreciate the full piece of music; you must tune into all instruments.

Turning to exoplanets (planets around other stars): we have discovered over 4,000 planets around other star systems and we look at their light in great detail to understand them. One very useful instrument is a spectroscope which takes the light from the object and splits it into its component parts. When we do that, we can explore a feature known as *absorption lines*.

Any time light travels through a gas, the gas strips out a little of the light. The mechanism for this is that gas is composed of atoms and atoms are made up of neutrons and protons in the nucleus and electrons whizzing around the outside. If an electron absorbs a little energy from the light passing through it then it jumps to a higher energy state. To visualise that, imagine Mercury travelling around the Sun, gaining energy and jumping to the orbit

of Venus which is further out. Studying light from exoplanets that has passed through their atmosphere reveals absorption lines and, as luck would have it, absorption lines are like a fingerprint so we can tell exactly what gasses are present to help us understand what an exoplanet atmosphere is composed of.

A great example of how this has helped us to probe the secrets of exoplanets is in orbit around a red dwarf star in the constellation of Leo, 120 light years away. The atmosphere of the planet, named K2-18b, has been observed and spectroscopic study has revealed signs of water vapour and methane. Subsequent observations by the James Webb Space Telescope (JWST) have revealed carbon dioxide, suggesting an atmosphere rich in hydrogen and perhaps a water ocean. A molecule known as dimethyl sulphide was also detected and tantalisingly this is a molecule which on Earth is only created by life!

Comparison of the size of exoplanets orbiting Kepler-37 to Mercury, Mars and Earth (NASA/Ames/JPL-Caltech)

FACT 3

Rogue planets wander freely through the Galaxy

Across the universe, the forces of gravity and celestial dynamics determine how objects move. When it comes to planets most people envisage them silently orbiting a host star, yet there is a strange class of planet – 'rogue planets'. As their name suggests, they do not tend to obey all the usual rules and instead of orbiting a star, they wander through space, orphaned from any parent star. Their origin is still a bit of a mystery, and quite how many are drifting through space is not known.

When it comes to 'normal' planets like Earth, they develop out of the vast cloud of gas and dust from which the host star forms. In the case of our own solar system, this happened about 4.5 billion years ago when the Sun formed and, as it evolved, a protoplanetary disc formed. Eventually, gravity caused the disc to collapse into the family of planets. When it comes to rogue planets, they either form and are ejected out of the system as a result of gravitational interactions with nearby stars, or they form completely independently.

To date, about 70 rogue planets have been discovered. Some astronomers suggest there could be more rogue planets in our Milk Way galaxy than stars. They come in a wide range of sizes, from Earth-sized to gas giants, and with ranging atmospheric composition. Despite their (sometimes) large size, they are difficult to detect due chiefly to their lack of a host star. There are a number of techniques used to detect planets around other stars and they all, in some way or another, rely upon observing the host star. Rogue planets however are void of a host, so are far more difficult to detect.

The lack of a host star makes it difficult to locate rogue planets but it also means that there really is no guide as to where to look. If you want to find exoplanets then you look around stars. To find rogue planets, well, you can look anywhere you like. The space between the stars is the best bet but that means you really are looking for the proverbial needle in a haystack. Of course it may just be possible that a rogue planet is captured as it wanders a little too close to a passing star.

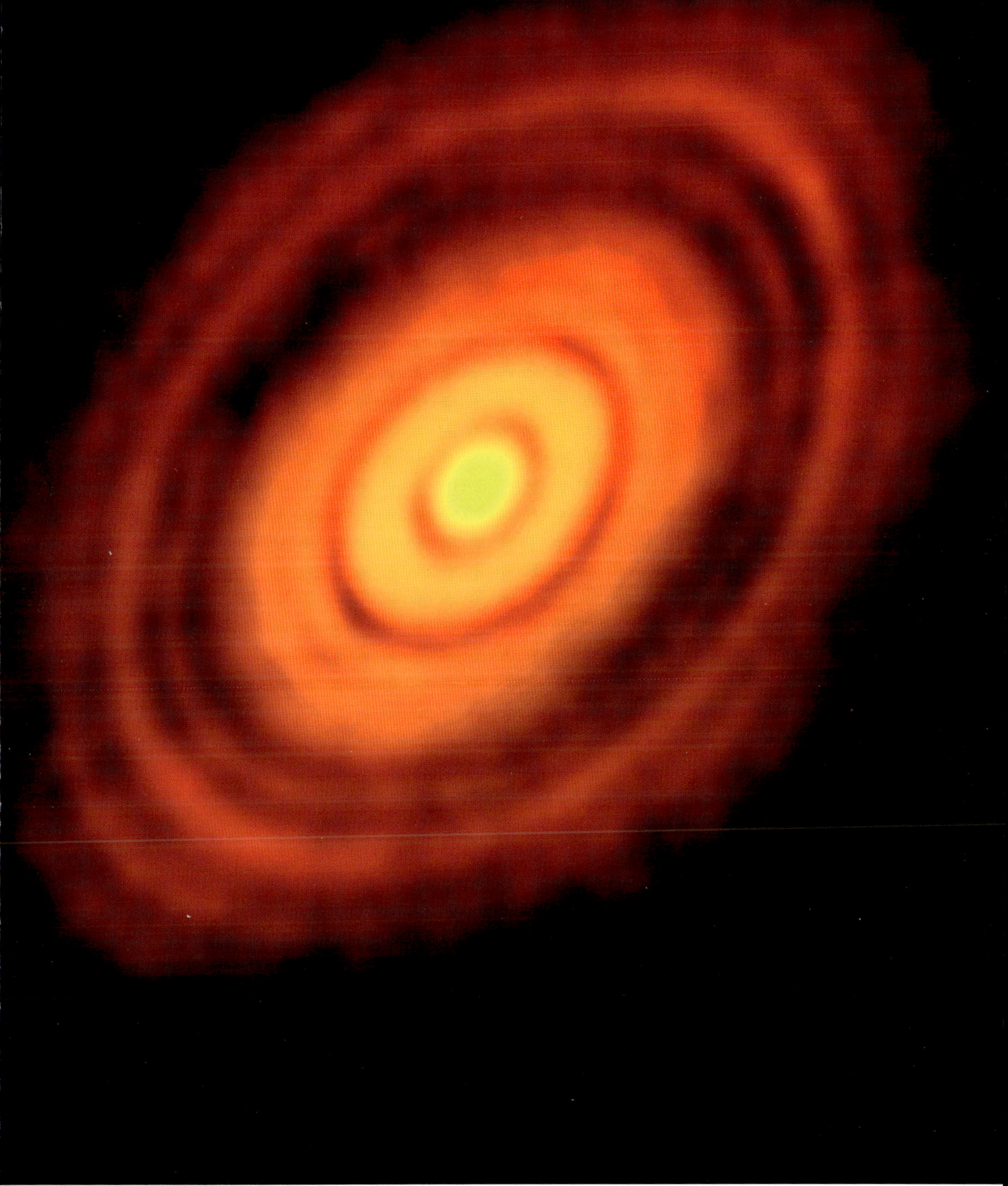

It is believed that most rogue planets form out of protoplanetary disks like this one captured around the star HL Tauri. (ALMA)

FACT 4

Gamma-ray bursts can release more energy in 10 seconds than our Sun will in its entire life

Gamma-ray bursts (GRBs) are among the most energetic events in the universe. As there name suggests, they are short-lived bursts of gamma radiation that can last from a few fractions of a second to several minutes. Even supernovae, the explosions that mark the death of massive stars, are outshined several hundred times more by a GRB.

Until a few years ago, we really did not understand what a GRB actually was. GRBs have remained a mystery to us since their discovery in the second half of the 1960s. This discovery, by US military satellites, was quite unexpected. They were on the lookout for nuclear testing by the then Soviet Union by detecting bursts of gamma radiation from the tests. Instead of detecting test nuclear explosions (which they found none of) they did detect events in the sky that were found to be coming from beyond our solar system.

Since this discovery there has of course been a lot of research into the origin of GRBs and there are now satellites like Swift and Fermi constantly scouring

Positions in the sky of all gamma-ray bursts detected during the BATSE mission. (NASA)

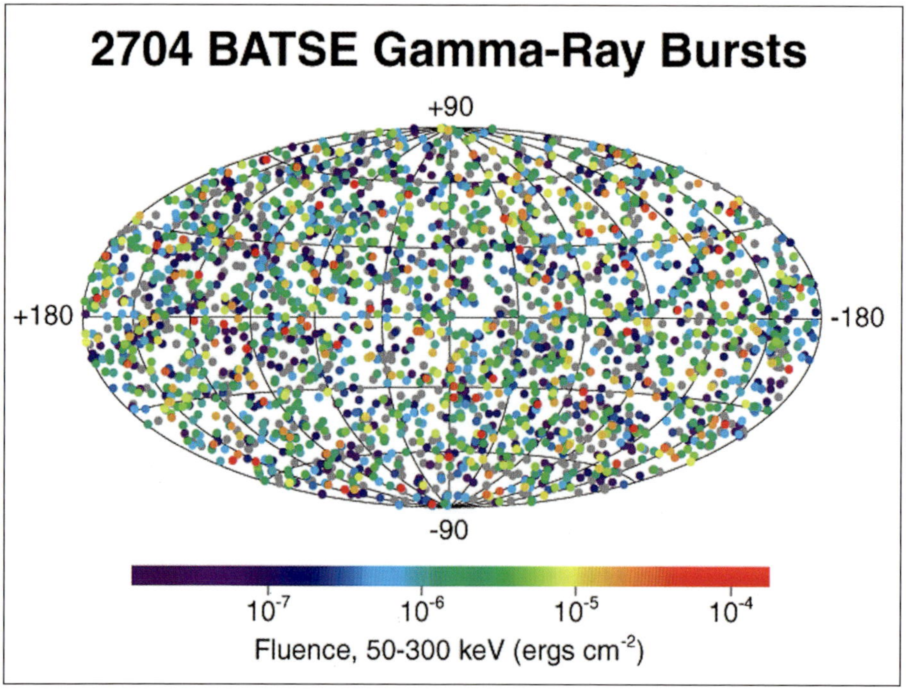

the sky for them. The studies have revealed that there seem to be two distinct classes of GRB, and both are related to the formation of black holes. Bursts that last from between two seconds and a few minutes are classed as long duration and they are associated with the death of a massive star during the supernova process. Note that not all supernova explosions generate GRBs but all GRBs of long duration are from supernovae. The short duration bursts last less than two seconds and seem to be the result of the merger of two neutron stars into a new black hole.

Neutron stars, GRBs, black holes and supernova explosions are all events relating to the lives of massive stars, at least eight times more massive than our Sun. The source of energy production in the Sun is slightly different and relies upon nuclear fusion. Deep in the core of the Sun (and all stars for that matter), the pressure is so intense that atoms are being smashed together. In the case of the Sun, this is predominantly hydrogen atoms being fused together to produce helium atoms.

The by-product of all of this is the release of energy. Taken over the entire mass of the Sun and over its entire lifetime, which will be of the order of 10 billion years, the overall energy released by the fusion process is in the region of 1.3×10^{44} which is 130 followed by 42 zeros or, if I were to make an attempt to put words around it, is 130 tredecillion joules! This does seem a lot but it is spread out over 10 billion years. GRBs on the other hand release a frankly unimaginable amount of energy and can put out as much energy in 10 seconds as the Sun across its entire life!

FACT 5

Some galaxies are so far away that we will never be able to see them

One of the greatest discoveries of the twentieth century was the expansion of the universe. Edwin Hubble (1889–1953) was working at the California Institute of Technology (Caltech), and during observations of distant galaxies he noticed a pattern: that all the galaxies seemed to be moving away from us, but more intriguingly that the further away they were, the faster they were travelling! This observation led to the development of Hubble's Law, which shows that the universe is expanding, and also helps us to pinpoint its age.

The concept of the light from objects in space taking a long time to reach us is really quite a bizarre one. Our experience here on Earth is somewhat different. After all, the distances are considerably less. When you turn the light on in your lounge, the light only has to travel a few metres to strike an object, bounce off and reach your eyes. Given that light travels around 300,000 kilometres per second its clear to see that it takes the tiniest fraction of a second for this to happen, so the time delay is unnoticeable. In space, the distances are, pardon the pun, astronomical. Light from the Sun takes just over eight minutes to reach us, the Moon 1.3 seconds, the nearest star about four years and the Andromeda Galaxy 2 million years.

The work of Edwin Hubble has shown us that the universe is 13.7 billion years old. The age of the universe and the finite speed of light means we experience a phenomenon known as the *observable universe*: a spherical region of space made up of all the matter in the universe that can be seen and observed from Earth. Due to the age of the universe, the radiation from all these objects has had time to reach us. Estimates for the number of galaxies in the observable universe range from several hundred billion to 2 trillion!

It is thought that the distance to the edge of the observable universe (current estimate is 46.5 billion light years) is the same in every direction. The question that of course pops up is what is beyond it. Well, the unobservable universe! Nothing can travel faster than the speed of light so there is a maximum distance, known as the *particle horizon*, beyond which objects will not be detectable. The time taken for light to reach us from galaxies beyond the particle horizon is greater than the time since the Big Bang so there has not been enough time for the signals to reach us and never will be. They will remain forever beyond our view.

Even those objects that we can currently see will eventually be lost. The universe is expanding at an accelerated rate so all objects beyond our local supercluster will eventually appear frozen in time to us as they approach the *proton horizon*. They will then become more and more red and slowly fade from view, never to be seen again.

The Hubble Space Telescope captured this 'Ultra Deep Field' revealing about 10,000 galaxies. Within them is UDFj-39546284 which lies at a distance of over 13 billion light years. There are other galaxies too far away to be observable. (NASA and ESA)

FACT 6

Most galaxies in the universe are dwarf galaxies

Galaxies were first discovered in the eighteenth century by Charles Messier (1730--1817), but at the time he did not realise it! Messier was a comet hunter and as he scoured the sky he came across other fuzzy blobs. Observing the objects over many nights he realised they were not moving so could not possibly be comets. Either way, although he was not too interested in them, he did make a catalogue record, so that he would not mistake them for comets again. Some of the fuzzy blobs Messier discovered were external galaxies.

Messier and other observers could tell that some of the objects had a spiral structure but it was not until the 1920s when Edwin Hubble (1889–1953) measured the distance to one of the spiral nebula discovered by Messier that the immense distances were discovered, confirming they were external galaxies.

Just over a hundred years on, we now know that galaxies are not just spiral in structure: there are elliptical galaxies, barred spiral galaxies, irregular galaxies and even dwarf galaxies. With the discovery of so many types, Hubble suggested that they could be grouped into certain classifications based on their shape. The Hubble Classification (or 'Hubble sequence') divides galaxies into four main types – spiral, elliptical, lenticular and irregular – and all galaxies that have been identified fall into those categories or one of the many sub-categories. The sub-categories are a bit of a mixed bag but include the dwarf galaxies, tiny in comparison to the others but, it seems, really quite numerous.

That the vast majority of galaxies are clustered together into groups. Our own galactic family is known as the *Local Group*, and includes all the galaxies in our region of space out to a distance of about 10 million light years. Our own Galaxy, the Milky Way, and the Andromeda Galaxy, are among the largest galaxies in our Local Group, which comprises at least 80 member galaxies, but contains well over 50 smaller dwarf galaxies.

If we consider the quantity of dwarf galaxies in our group then it is clear that spiral or elliptical galaxies are vastly outnumbered. Our galaxy alone is orbited by two well-known dwarf galaxies, the Large and Small Magellanic Clouds (LMC and SMC, visible only from the southern hemisphere) and an estimated 18 additional ones.

Many of the dwarf galaxies seen in the universe today are thought to have been ripped off some of the larger galaxies as a result of a gravitational interaction with another galaxy. While many of them originate from larger galaxies, a great proportion will become at one with their larger cousins too. Immense galaxies have an enormous gravitational pull so any dwarf galaxies that wander too close will easily be sucked in and absorbed by the larger galaxy. The nature of dwarf galaxies may be more transitory than larger galaxies but, whether they come or whether they go, they are still far more numerous across the cosmos.

Dwarf galaxies like NGC5264 typically possess around a billion stars. (ESA/Hubble)

FACT 7

Stars do not actually twinkle

A well-known nursery rhyme that children all across the world are familiar with starts, 'Twinkle twinkle little star, how I wonder what you are?'. We now know far more about the stars than ever before and one thing we have learned is that, contrary to popular belief, stars do not actually twinkle.

Stars are formed out of vast molecular clouds of gas and dust and, over time, gravity causes the cloud to collapse causing localised concentrations of

There are about 6,000 stars visible to the naked eye butu there are countless more in the universe. Despite what your eyes tell you, stars don't actually twinkllel, eveen those captured in this immage from the Hubble Space Telescope from the star cluster known as Liller 1. (ESA/Hubble & NASA, F. Ferraro)

material. As more and more matter collects the collective mass slowly increases as does the force of gravity it exerts. Deep inside the core of this *protostar*, the pressures and temperatures reach a critical level, fusion begins and the star begins to shine. In the case of most stars, this is caused by the fusion of hydrogen atoms.

For the most part of a star's life the nuclear fusion in the core creates an outward force known as *thermonuclear pressure*. This force is balanced by the inward pull of gravity and for millions of years, often billions, the star remains stable. As the star evolves, the various forces increase and decrease at different times causing the star to increase and decrease in size and brightness. Generally, these changes take place over millions of years and certainly cannot be responsible for the twinkling effect.

There are other causes for star brightness to change – an orbiting companion or the passage of large 'starspots' across the surface of a star. There are even some stars (in fact a great proportion of the stars you can see in the sky) called *variable stars* which, for one reason or another, vary in their output of light, but this happens over hours and days, occasionally minutes, but never fast enough to give a twinkling effect. To understand the twinkling effect we need to look much closer to home.

Whatever the type of star, their light travels across great voids of space to reach us. When it reaches us here on Earth, imagine it as a thin beam of light. As it travels through our atmosphere, the shell of gas that surrounds our planet, any movement of the gas will knock the incoming beam of light around, giving the illusion of flickering. The effect is far more pronounced when stars are lower in the sky because the light has to travel through more of the atmosphere. The weather conditions can greatly effect the stability of the atmosphere too, which means some nights the stars seem to twinkle more than others.

FACT 8

The universe is expanding at an accelerated rate

Since the early 1900s we have known that the universe is expanding. Edwin Hubble (1889–1953) was the first to provide real evidence for this when he closely studied the light from distant galaxies. Using spectroscopic techniques to split incoming light from galaxies into their component parts (much like water drops split sunlight into a rainbow), Hubble discovered that the characteristic dark lines in the spectrum of the galaxies were shifted toward the red end. These dark lines are produced by the presence of gas and, all things being equal, they should always appear in the same place. What Hubble found was that the lines were *redshifted* due to the galaxies all moving away from us, due to the expansion of the universe.

Studying the shift of the lines meant we could calculate the velocity of a given galaxy and Hubble realised there was a correlation between the distance to the galaxy and its velocity. Knowing the distance to galaxies and knowing how fast they were moving meant we could, for the first time, calculate at what point in time all the galaxies occupied the same region of space. This marked the 'Big Bang' and the popping into existence of the universe. Doing the maths and using some modern data to support the evidence reveals that this moment occurred 13.7 billion years ago.

The question on everyone's mind that was an obvious step, was what will happen in the future? Will the universe continue expanding forever? The answer lies in the amount of matter in the universe: if there is enough matter with sufficient gravity then the expansion may one day slow, halt and reverse; or if there is insufficient mass and therefore insufficient gravity to halt the expansion, it may just continue to expand forever. If we look at all the mass in the universe, including the dark matter, then we are way short, suggesting the universe will continue to expand forever.

Hold that thought. Enter the observations of supernovae. Supernovae are massive stars that have exploded at the end of their lives. Fortunately, due to the processes that drive a supernova, their actual luminosity is consistent regardless of their location, so we can use them as standard markers. Knowing how bright a supernova *should be*, from analysis of its spectrum, we can observe how bright it *appears* in the sky and therefore deduce, with accuracy, its location. If we do this for enough supernovae we might be able to validate the current size estimates of the universe.

Surprisingly perhaps, it turns out that the supernovae observations reveal the universe is bigger now than we first thought. Further analysis reveals that the universe is expanding at an accelerated rate rather than slowing down or just continuing! An accelerated expansion of the universe was not expected but is an intriguing observation that forces us to review our understanding of the basics of the universe. Accelerated expansion can be measured. We now need

The Crab Nebula is an example of a supernova remnant. The study of supernova has led to the discovery of the expanding universe. (NASA/ESA)

to understand what is driving this expansion. One of the most popular theories is *dark energy*. Tiny subatomic particles popping in and out of existence might just produce enough of a pushing force to be driving the accelerated expansion. For now, it is just a theory but it is the only one we have, and further observations will be required to further analyse the accelerated expansion of the universe.

FACT 9

Water is common in space

Water in space represents a fascinating and tantalising concept. If water is present, it not only has an impact on our ability to travel further in space but also on our understanding of the prevalence of life in the cosmos. In its various states (gas, liquid and solid), water it seems is fairly common across the universe from the familiar oceans on Earth, to the subterranean global oceans on Jupiter's moon Europa, and from the dark, cold craters on the Moon to gigantic quantities of water in molecular form in gas clouds in space. These great reservoirs of water could be used by space travellers for sustenance or the hydrogen and oxygen components could be processed for rocket fuel to enhance the reach of our species out into space.

We are familiar with water on Earth but it has not always been here. It is thought that the water on Earth was brought here by comets. These relatively small icy bodies that orbit in the cold depths of the solar system collided with the young Earth billions of years ago and delivered molecules of water to our planet. Studies over recent decades have revealed water is not just present on Earth but has been found in all manner of places in space. Closer to home, there are craters in the polar region of the Moon where the incoming sunlight strikes at such a low angle that the crater bottoms are kept constantly in the

Water detected on exoplanet WASP-96b based on James Webb Space Telescope. (JWST) spectral studies. (NASA/ESA/CSA/STScI)

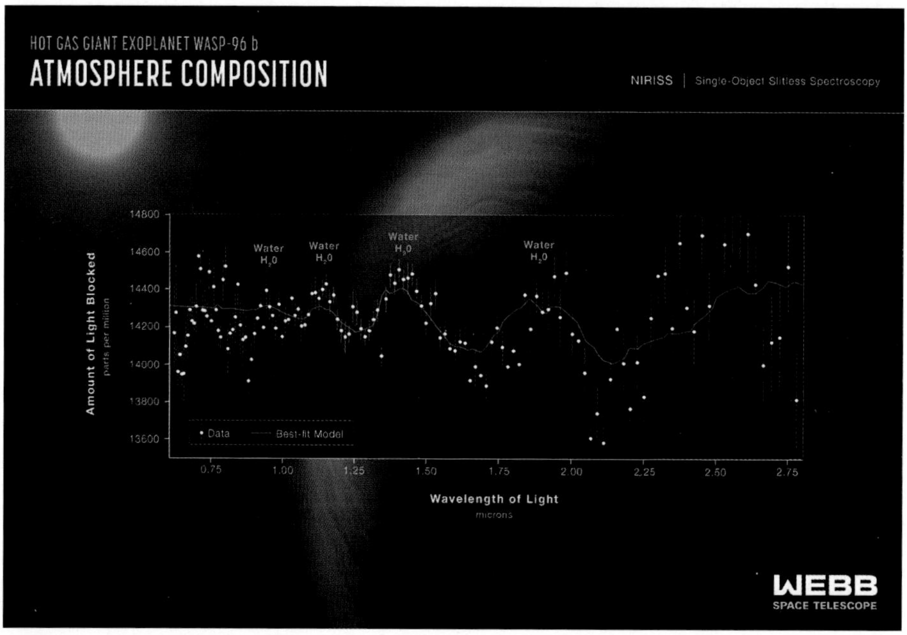

dark. At these dark, cold places, ice has accumulated so much that there is an estimated 600 billion kilogrammes of water ice present.

Further out in the solar system there is water ice on the surface of Mars and on a number of asteroids, but it is around some of the gas planets where things get really interesting. Jupiter has 95 moons but one in particular, Europa, is of great interest. Europa is an icy moon that is subjected to intense tidal forces from Jupiter. The forces constantly stress the interior of the moon, generating heat which has melted the lower levels of ice to form a giant, global ocean under the icy crust. This brings with it the possibility of life at the bottom of the ocean, just as entire ecosystems exist at the bottom of our oceans.

Beyond our solar system there is even water in the depths of space. Water is made of hydrogen and oxygen. Hydrogen is the most common element in the universe and oxygen is made through fusion in the core of stars and dispersed when the stars die through supernova explosions. The two elements mix in gigantic gas clouds to form water molecules. Due to the conditions in space, 99% of the water in clouds is ice condensed onto dust grains and the remaining 1% is gas.

It will be some time before we can make any use of water in distant gas clouds but it does lead to interesting debates about the prevalence of water in deep space and on exoplanets. Water is common in our solar system and seems to be common in space. It is therefore reasonable to assume water is common on exoplanets – planets around distant stars. If so, then this leads us to the conclusion that the conditions for life may be common in the universe and perhaps life is even more common than we thought.

Water on the Moon is more likely to be of use as we step further out into the solar system. NASA's Artemis Program is the first step to establishing a permanent lunar base and the presence of water will be of immense use. Once established, the water can be harvested to serve as drinking water to sustain the inhabitants and provide fuel for power systems and propulsion.

FACT 10

The first exoplanet was discovered in 1992

Until 1992, humanity had been under the assumption that the only planets in the universe were those in orbit around our Sun. In other words, the planets of our solar system. This changed when Aleksander Wolszczan and Dale Frail announced the discovery of two planets orbiting around a pulsar named PSR B1 257+12 (pulsars are the remains of massive stars that have reached the ends of their lives).

Exoplanets are defined in a slightly different way to the planets in our solar system and they must meet three criteria to be classed as an exoplanet:

They must be orbiting around a star or stellar remnant *and* have a mass low enough that the fusion of hydrogen has not begun in the core.

Anything above the mass in item 1 should be classed as a 'brown dwarf' regardless of its location.

Anything that is a 'free floating' object in a star cluster and with the mass low enough for hydrogen fusion to be able to commence is classed as a 'sub-brown dwarf'.

The discovery of the first exoplanets fired a new search for more, and several techniques have been developed that do not rely upon direct observations, although some exoplanets have now been imaged directly. Perhaps the most common methods are the *transit* and the *radial velocity* methods.

The *transit method* involves observing the light from a star and measuring its brightness. As a planet passes in between us and the star in an event known as a transit, the light from the star will dip a little. The amount of reduction in light received varies based on the size of the planet and the star, but to demonstrate the challenge, HD 209458 in Pegasus dims by just 1.7% as the planet passes in front of the star. Many dim by even less than this: for an Earth-sized planet transiting a star like the Sun, the light reduction will be around 0.008%! Measuring the light from stars and plotting their brightness on a graph with brightness on the Y axis and time on the X axis allows us to show the light curve of the system and from this we can determine the presence and size of exoplanets transiting the star.

The *radial velocity method* is different. Any star with a planet in orbit around it will be very slightly 'wobbled' from its usual position. We can detect this wobble by studying the spectrum of distant stars, which reveals a tiny shift in characteristic lines in the stellar spectrum. As the star moves toward us, the lines shift to the blue end, but as it moves away they shift to the red end. Measuring the amount of this shift tells us how much the star is wobbling due to the presence of an exoplanet and from this we can even deduce the mass of the unseen object.

There are other techniques that have enabled us to detect exoplanets and, at the time of writing, over 5,200 exoplanets have been discovered. That is 5,000 planets orbiting around other stars in our Galaxy! It seems planetary formation is common, at least in this part of the universe, and if it is then there is no reason to suspect it is not common in the universe as a whole.

Artist's impression of the planets orbiting PSR B1257+12. The one in the foreground is planet "C". (NASA/JPL/Caltech)

FACT 11

We have seen a galaxy whose light travelled 33 billion years to reach us

It is truly wonderful to witness the awe on people's faces when you explain that observing distant celestial bodies means you are peering back into the depths of time itself. By extending our gaze to the farthest reaches of the cosmos, we can observe objects so remote that their light left them aeons before the birth of our own solar system. With the advent of the James Webb Space Telescope (JWST) the race was on to find the most remote galaxies in existence, perhaps even those that heralded the start of galactic formations.

Launched aboard an Ariane 5 rocket in December 2021, the JWST began its celestial voyage, ultimately positioning itself 1.5 million kilometres from Earth. From there it began its work. Among its myriad tasks, the telescope turned its

The galaxy Z13 is smaller in size than the Milky Way but one of the most distant objects ever observed. (NASA/CSA/ESA/STScI)

gaze towards Pandora's Cluster in the Sculptor constellation, utilising its near-infrared (NIR) camera to capture individual exposures lasting four to six hours but totalling an impressive 30 hours of data accumulation: the restful, 'deep-field' image of the region.

The selection of this particular area of the sky was deliberate, owing to the presence of several galaxy clusters capable of *gravitational lensing*. This phenomenon naturally magnifies distant objects, giving astronomers an extra edge. The resultant image unveiled a staggering array of 60,000 celestial bodies, meticulously whittled down to a mere 700, from which a final selection of eight galaxies emerged. These were believed to be among the earliest galaxies to have evolved across the entire universe.

However, among the eight galaxies, astronomers fell short of the ultimate prize and instead detected the second most distant galaxy. It was named Uncover Z-13 and lay at an astonishing distance of 33 billion light years away. This assertion may initially perplex some, given our understanding that the universe is approximately 13.77 billion years old. However, it is important to understand that the light from Z-13 began its journey when the universe was a mere 332 million years old. Over the billions of years that followed and as the universe expanded, it carried this distant galaxy along with it, stretching its estimated distance to an incredible 33 billion light years.

The research team uncovered another galaxy (Uncover Z-12), ranking fourth in the distance hierarchy. Yet what was not expected was that both galaxies were unexpectedly large in comparison to other galaxies at such distances. It is important to note that their scale, while remarkable in the context of the early universe, pales into insignificance when compared to the galaxies we observe today. Z-13 (and its counterpart, Uncover Z-12) span approximately 2,000 light years, whereas the Milky Way sprawls across roughly 100,000 light years.

During the early stages of the universe, galaxies were substantially smaller and more densely packed. Consequently, at these immense distances, galaxies typically manifest as faint red specks in images. However, due to their comparatively larger size, these two entities appear akin to soft, diffuse spheres.

Delving into the intricacies of galaxies at such remote distances gives us a unique vantage point to unravel the mysteries of the early universe and glean insights into its evolutionary path. It is not just the galaxies we can learn about. The light from them has travelled through the universe and through the rarefied gas that made up the early cosmos. By analysing their spectral signatures, we can start to learn more about the elements and underlying physics that governed the universe in the aftermath of the Big Bang.

FACT 12

We have successfully redirected an asteroid

Deep Impact and *Armageddon* were both fabulous films. Packed with action and the imminent threat of an asteroid hurtling towards Earth, what's not to love? While both movies touched on our potential strategies to avert such a calamity, the reality is far removed from Hollywood's dramatisations.

On 26 September 2022, and without the help of Bruce Willis, we did achieve a historic milestone in planetary defence when NASA's Double Asteroid Redirection Test (DART) spacecraft successfully collided with the asteroid moonlet Dimorphos. This groundbreaking mission marked the first time

High resolution image of Dimorphos created by combining the final 10 full-frame images obtained by DART's Didymos Reconnaissance and Asteroid Camera for Optical Navigation. (NASA/Johns Hopkins)

humans had intentionally altered the motion of a celestial body, demonstrating our capability to protect Earth from potential asteroid threats.

DART targeted Dimorphos, a small moonlet approximately 160 metres in diameter that orbits the larger asteroid Didymos. This binary asteroid system was chosen because it posed no threat to Earth, making it an ideal testing ground. The mission aimed to change Dimorphos's orbital period around Didymos by at least 73 seconds through kinetic impact.

The spacecraft, roughly the size of a refrigerator and weighing 610 kilograms, relied on kinetic impact technology – in other words, on becoming a high-speed projectile. Travelling at approximately 6.14 kilometres per second, DART struck Dimorphos head on, transferring momentum to alter the moonlet's trajectory. This technique represents the most technologically mature approach for deflecting potentially hazardous asteroids.

The mission exceeded all expectations. Instead of the minimum 73-second change, DART shortened Dimorphos's orbital period by an impressive 32 minutes. Ground-based telescopes and the Italian Space Agency's LICIA Cube, which separated from DART before impact, captured dramatic images of the collision and the resulting debris plume that extended thousands of kilometres into space.

The success of DART validates kinetic impact as a viable planetary defence strategy, particularly for smaller asteroids. However, the mission also highlighted the need for early detection systems, as this technique requires years of advance warning to be effective. For larger asteroids, alternative methods like gravity tractors or nuclear deflection might be necessary.

Fortunately, efforts to detect and track space rocks are ongoing and are accompanied by the development of defensive strategies. Nonetheless, the elusive nature of asteroids, often small and camouflaged in the vastness of space, poses a significant challenge akin to finding a cosmic needle in a haystack.

FACT 13

Early galaxies look nothing like what we see today

When discussing galaxies, the mind often drifts to visions of spirals or ellipticals. The galaxy we live in is known as the Milky Way and our solar system is about two thirds away from the centre. It is a great big spiral galaxy. To visualise it, think of two fried eggs stuck back to back, the yolks of the eggs representing the central nucleus of the galaxy and the whites of the eggs the spiral arms. Current estimates suggest the Milky Way contains somewhere in the region of 200 billion stars.

The concept of galaxies dates back to the eighteenth century when French astronomer Charles Messier (1730-–1817), the renowned comet hunter, stumbled upon numerous fuzzy celestial objects distinct from the comets he was hunting. He meticulously catalogued these findings in his now well-known catalogue of deep sky objects. However, it was not until the 1920s, when Edwin Hubble (1889–1953) measured the distances to some of these 'objects', that they were recognised as external galaxies.

The James Webb Space Telescope (JWST) has been hunting galaxies that existed when the universe was young. One survey reveals that the early universe was teeming with galaxies of diverse shapes. In the initial 6 billion years of cosmic evolution, up to 80% of galaxies had a strikingly flat appearance, looking rather like surfboards. Yet, alongside these were galaxies resembling swimming pool noodles, adding to the mix. The reasons behind their different shapes in this epoch remains a mystery.

Examining imagery from the telescope's Cosmic Evolution Early Release Science (CEERS) Survey, researchers concentrated on galaxies dating back to a period when the universe was between 600 million and 6 billion years old. To their surprise, the expected mix of spiral, elliptical and irregular galaxies was notably absent. Instead, they encountered a majority – ranging from 50 to 80% – displaying a flattened, two-dimensional profile.

Among these newfound shapes were more than just surfboard- and pool-noodle-looking galaxies but some looking like frisbees and spheres. Surprisingly, the surfboard shape predominated, with spherical galaxies being the most rare. Additionally, there was an increase in the prevalence of frisbee- and noodle-shaped galaxies as the era approached 6 billion years.

The early galaxies exhibited significantly lower mass compared to their modern counterparts. Their limited growth period, when compared with today's more massive galaxies, hints at quite a stark contrast in cosmic evolution. The research owed much to the JWST's remarkable sensitivity and resolution, enabling a delve into this incredible cosmic archaeology.

Our own Milky Way has also been studied and it is thought that, if we could go back in time, its appearance would change. In our hypothetical journey back

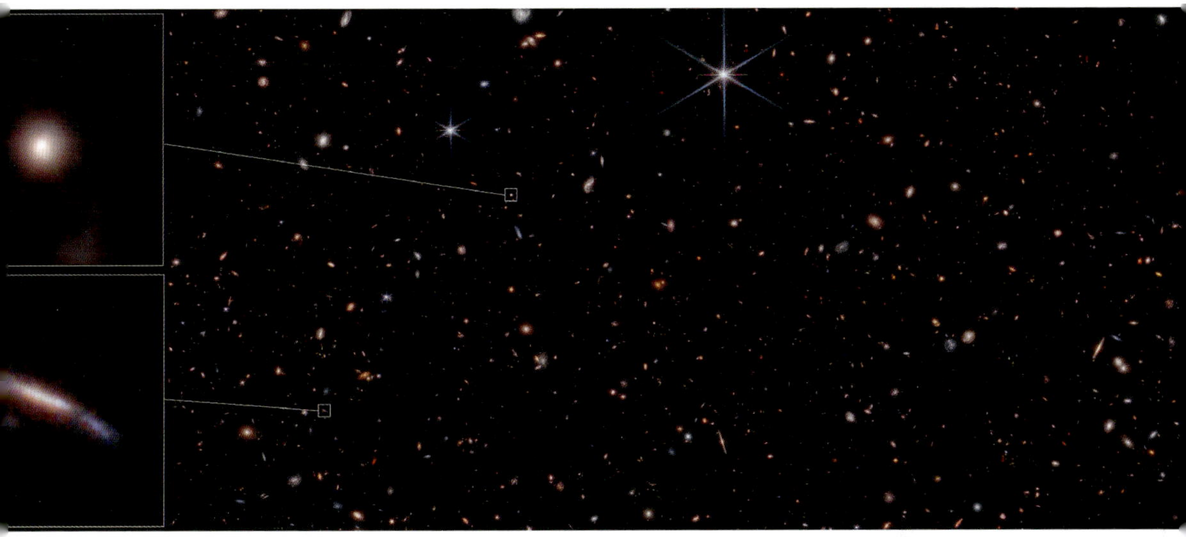

Early galaxies captured by the James Webb Space Telescope's CEERS Survey, showing the unusual shapes — including spheres and elongated forms — that dominated the universe between 600 million and 6 billion years after the Big Bang. (NASA/ESA/CSA/STScI)

in time, it is likely that the Milky Way would resemble a surfboard shape, rather different to the familiar spiral structure we recognise today.

Despite our accumulated knowledge of galaxies over the years, and aided significantly by imagery from the Hubble Space Telescope, the insights from the JWST underscore the depth of our ignorance. As we continue to explore the cosmos, delving deeper into space and farther back in time, we are merely scratching the surface of the galaxies that inhabit it.

FACT 14

There is very little dark matter at the core of the Milky Way

The ever-evolving landscape of science keeps us on our toes, constantly challenging our understanding of the universe. Our current view of the universe is that matter came before galaxies, and this is quite an intuitive view. But the motion of stars within these galaxies has puzzled astronomers – stars further from the centre seemed to move slower than anticipated. This discrepancy prompted the suggestion that maybe *dark matter* was the cause.

Dark matter remains an enigma, exhibiting behaviour unlike anything observed in the realm of ordinary matter. Devoid of luminosity, it neither emits, absorbs nor reflects light or any form of radiation, rendering it virtually invisible – a fitting epithet. Despite comprising an estimated 27% of the universe, dark matter has still eluded our direct detection, its presence inferred solely through its gravitational interactions with light and ordinary matter.

Researchers at the Massachusetts Institute of Technology (MIT) delved into the mysteries of the Milky Way, meticulously measuring the velocities of stars across its expanse. Surprisingly, and as noted earlier, the stars in the outer reaches were found to orbit at a lower velocity than expected. This hinted at a core less dense in mass than previously presumed and, consequently, potentially containing less dark matter. It is possible to estimate the amount of ordinary matter based on observation, but taking this into account, the core must have some unseen matter, and it has been thought that this would likely be dark matter.

Using data from the *Gaia* satellite and the Apache Point Observatory Galactic Evolution Experiment (APOGEE), rotation curves were constructed charting the velocities of stars against their distance from the centre of the galaxy. These meticulous observations extended to over 33,000 stars, spanning distances of up to 30 kiloparsecs (where one parsec is the equivalent of 3.26 light years). Doing this enabled researchers to estimate the amounts and distribution of dark matter within galaxies.

The resultant curve defied expectations, revealing a steeper decline in velocity over distance than previously assumed. Stars situated farther from the galactic core exhibited slower motion, suggesting a lesser concentration of matter at the centre. This revised understanding implies that the observable, baryonic matter may require less dark matter to account for observed phenomena.

Typically, we have only attempted to measure the rotation in external galaxies, but thanks to APOGEE and *Gaia* the Milky Way became a target. Ever since the 1970s we had thought that there was significant amounts of dark matter at the core but this new observation contradicts this theory.

To get a full grasp of the situation and whether dark matter is more or less prevalent in galactic centres, further exploration of galaxies like the Milky Way is key. This promises to reshape our comprehension of the universe's dark matter content and, as the quest for knowledge continues, each revelation nudges us closer to unravelling the cosmic mysteries.

Artist's impression of ESA's *Gaia* satellite observing the Milky Way. (ESA/ATG)

FACT 15

The Kuiper belt of icy objects is bigger than we thought

Beyond the familiar orbits of the major planets lies a vast cloud of objects known as the Kuiper belt. It extends outward from the orbit of Neptune for what was originally thought to be approximately 8 billion kilometres. This theory of such a belt existing was first suggested by astronomer Gerard Kuiper (1905–1973) in the mid twentieth century, and it is thought to be home to billions of icy bodies and dwarf planets. Understanding its origin means we may start to understand more about the origins of our solar system.

Artist's impression of the *New Horizons* spacecraft on its way to the Kuiper belt via Pluto. (NASA)

Of all the space missions over recent years NASA's *New Horizons* spacecraft has been one that has delivered some stunning insights into the worlds at the edge of our solar system. Launched atop an Atlas V rocket in January 2006, *New Horizons* began its historic journey to explore the outer reaches of our solar system. Its primary destination was the mysterious dwarf planet Pluto, which it reached after an incredible 9.5-year journey.

The successful fly-by of Pluto in 2015 marked a monumental achievement for planetary science and space exploration, providing unprecedented insights into this distant world and its icy companions. Yet, *New Horizons* did not stop there. With Pluto behind it, the spacecraft set its sights on the next frontier: the Kuiper belt itself and the edge of the solar system.

Since entering the Kuiper belt, *New Horizons* has been silently drifting across the cold depths of space at a distance nearly 60 times the distance from Earth to the Sun. En route, its Venetia Burney Student Dust Counter (VBSDC) (a remarkable instrument crafted by students at the Laboratory for Atmospheric Space Physics at the University of Colorado, Boulder) diligently monitors dust levels. These microscopic particles, borne of collisions between Kuiper belt objects (KBOs), offer valuable insights into the dynamics of the depths of the solar system.

Some of the dust particle detections recorded by the VBSDC suggest there have been many collisions and interactions among KBOs. The readings, and their distribution as *New Horizons* slips through the belt suggest that this may be more extensive than thought or perhaps that there may be a second belt altogether.

The data underpinning these discoveries were collected over a three-year period as *New Horizons* travelled through the region between 45 and 55 astronomical units from the Sun, where one astronomical unit represents the average distance between Earth and the Sun. At the same time, ground-based observatories, such as the 8.2-metre optical-infrared Subaru Telescope in Hawaii, have been instrumental in expanding our catalogue of KBOs, further revealing the shear scale of the Kuiper belt.

Together, these findings paint a new picture of a dynamic and complex region of our solar system. As *New Horizons* continues its journey through the Kuiper belt, its extended mission holds the promise of further revelations, potentially stretching into the 2040s and beyond.

FACT 16

Betelgeuse's surface is boiling furiously

Orion, the hunter, is one of the most recognisable constellations in the night sky. Named after the mythological Greek hunter, Orion is a very distinctive pattern of bright stars, forming the shape of the hunter holding a club and shield. It is well known as being home to the Orion Nebula but it also has some wonderfully bright stars, with Betelgeuse perhaps the most well known.

A red supergiant, whose name translates in some languages as 'armpit of the giant', Betelgeuse stands out as one of the most readily recognisable stars in the sky. It has a striking red colour easily noticeable to the naked eye and it sits at the upper left corner of the constellation of Orion. Any newcomers to the world of stargazing often learn the location of Betelgeuse and use it as a signpost to other celestial wonders. While all stars are considered to be immense, Betelgeuse dwarfs many with its incredible size of 1.2 billion kilometres across, nearly 2,000 times the diameter of our Sun.

Betelgeuse holds more surprises than just its sheer size. Recent observations have suggested an unexpected phenomenon: it spins with remarkable speed, defying conventional expectations for a star at its evolutionary stage. The Atacama Large Millimeter Array (ALMA), a radio interferometer comprising 66 antennae, has studied Betelgeuse's rotation. Analysis has revealed a strange pattern – while one hemisphere appeared to approach, the other seemed to recede, indicating a rotational velocity of 5 kilometres per second.

However, the story does not end there. Betelgeuse's surface is not a perfect sphere – it is characterised by quite violent convection processes within its photosphere (the visible surface), distorting its shape. The convection cells, some as large as the Earth's orbit around the Sun, churn at speeds exceeding 30 kilometres per second.

Researchers from the Max Planck Institute for Astrophysics have suggested that the observed velocity asymmetry might be the result not of rotational motion, but of the convective upheavals across Betelgeuse's surface. The theory was that the rise and fall of these immense convection cells could be behind the observed rotational anomaly.

To test this theory, an innovative processing technique was developed to generate synthetic data from ALMA observations. Surprisingly, in a majority of cases, the characteristic 'boiling' motion of convection cells remained elusive. The implication? Betelgeuse's rotational movement may be more complex than previously thought, with convection exerting a significant influence.

However, we are a long way from understanding Betelgeuse. Further observations, facilitated by instruments with a far greater resolution, are needed to unravel the true nature of this stellar giant. Encouragingly, recent high-resolution observations conducted in 2022 hold promise, although we still have to wait for the results of this study. As astronomers eagerly delve into the mysteries of Betelgeuse, one thing is for sure: there are plenty more surprises just waiting for us.

Image showing Betelgeuse (top left) and the dense nebulae of the Orion molecular cloud complex. (Rogelio Bernal Andreo)

FACT 17

Neutron stars are jetting material away at 40% the speed of light

It is widely understood that black holes possess quite the appetite, consuming anything in their way. Prior to disappearance into a black hole, matter often coalesces into swirling accretion discs. Similar to their progenitor stars, black holes boast formidable magnetic fields capable of generating powerful jets that erupt from their vicinity. Until recently, it was thought that only black holes generated these jets.

However, neutron stars, the dense remnants of massive stars, often similar in size to Earth, exhibit a comparable phenomenon, especially when in orbit around other stars. Recent observations have revealed jets emanating from neutron stars at velocities reaching an astonishing 35–40% of the speed of light.

The European Space Agency (ESA) commissioned the International Gamma-Ray Astrophysics Laboratory (INTEGRAL) in October 2002, with the primary objective of scrutinising gamma ray events throughout the universe, detecting energies up to 8 MeV (megaelectron volts). A gamma ray event is simply a burst of gamma radiation, and these short yet powerful bursts appear in an instant, making their origin difficult to track down. INTEGRAL's incredible sensitivity enables us not only to image gamma ray phenomena but also conduct a comprehensive spectroscopic analysis, making it invaluable for astrophysical research. Utilising INTEGRAL, astronomers identified the high velocity jets shooting from neutron stars in our Galaxy.

One of the primary challenges in determining jet velocities lies in tracking the movement of matter along their trajectories, given the immense distances involved. In a stroke of ingenuity, a team of astronomers proposed a novel approach utilising neutron star jets. Neutron stars can gravitationally strip material from companion stars, which, in a similar way to black holes, leads to the formation of accretion discs and subsequent ejection of matter in those powerful jets. The theory of these fascinating jets suggests that the energy emanating from the neutron star's surface can accelerate ejected matter, resulting in measurable disturbances that might facilitate easier tracking. With 125 neutron stars exhibiting similar behaviour, the prospect of observing numerous instances of jet ejections offers a promising avenue for understanding whether magnetic fields or other factors are at play.

Observations of two neutron stars, 4U 1728-34 and 4U 1636-536, gave great insight into jet formation. While both exhibited X-ray burst events, only 4U 1728-34 emitted observable radio waves, necessitating a joint effort from radio telescopes worldwide to capture supporting data. Surprisingly perhaps, and despite the violent nature of X-ray bursts, the locations from which jets launched remained intact, underscoring the resilience of these incredible stars.

A neutron star lies at the heart of the Crab Nebula in Taurus. (ESA/Hubble)

This novel approach demonstrates the feasibility of observing neutron star jets. Continued observations and analyses hold the promise of unravelling the intricate mechanisms governing jet formation and dynamics, shedding light on fundamental processes shaping our universe.

FACT 18

One teaspoon of neutron star material would weigh the same as 2.6 billion people

Look at the stars in the sky and you might be forgiven for thinking they are pretty static, glowing beacons that are never changing, casting their light across the cosmos. Well they certainly change – they form, fuse elements in their cores for billions of years, and then die. They move too, but the distances to the stars are so immense that viewed from Earth they seem to be static. It is no surprise then that we perceive them as unchanging.

During the main part of the life of a star, elements are fused in the core to generate the heat and light we perceive. In our Sun, which is just a normal star, hydrogen is being fused into helium. In doing so, a little matter is converted into heat and light. As the fusion progresses, an outward pressure is exerted, known as *thermonuclear force*. This is balanced by the force of gravity trying to collapse the star, leading to stability. As the star ages, it goes through different phases, where different elements are fused in the core, leading to the two forces increasing and decreasing. This leads to the star growing and shrinking in size over time.

Stars like our Sun will end up with a carbon-rich core and a powerful thermonuclear force that overcomes the force of gravity. Stars more massive than the Sun will have a stronger force of gravity which will be able to resist the increasing thermonuclear force. Eventually though, the core of the supermassive star will be rich in iron. It is not possible to fuse iron into any other element and for that reason, in an instant, the fusion stops, thermonuclear force ceases and the core collapses. Following the collapse, all the material that makes up the rest of the star crashes down on the collapsing core, compressing it to incredibly high densities. The material rebounds and the star explodes in an event known as a supernova.

What remains after an event like this is either a neutron star, a pulsar or a black hole. Which one is dependent on the mass of the progenitor star. They differ by way of mass and density, with black holes being the ultimate stellar corpse. Neutron stars are interesting in their own right though. Atoms are made of protons (positively charged), neutrons (no charge) and electrons (negative charge). When the core collapses and the material crushes down onto it, all the atoms are subject to such extreme forces that protons and electrons are combined to form neutrons and all the space between the neutrons is squeezed out. The result is a massive neutron 20 kilometres across.

If you could take a spoonful of neutrons from, for example, a neutron star then it would weigh around 10 million tonnes. That's equivalent to 2.6 billion people all standing on a set of scales together! If you cannot get your head round that number of people, then think of the population of China, and double it!

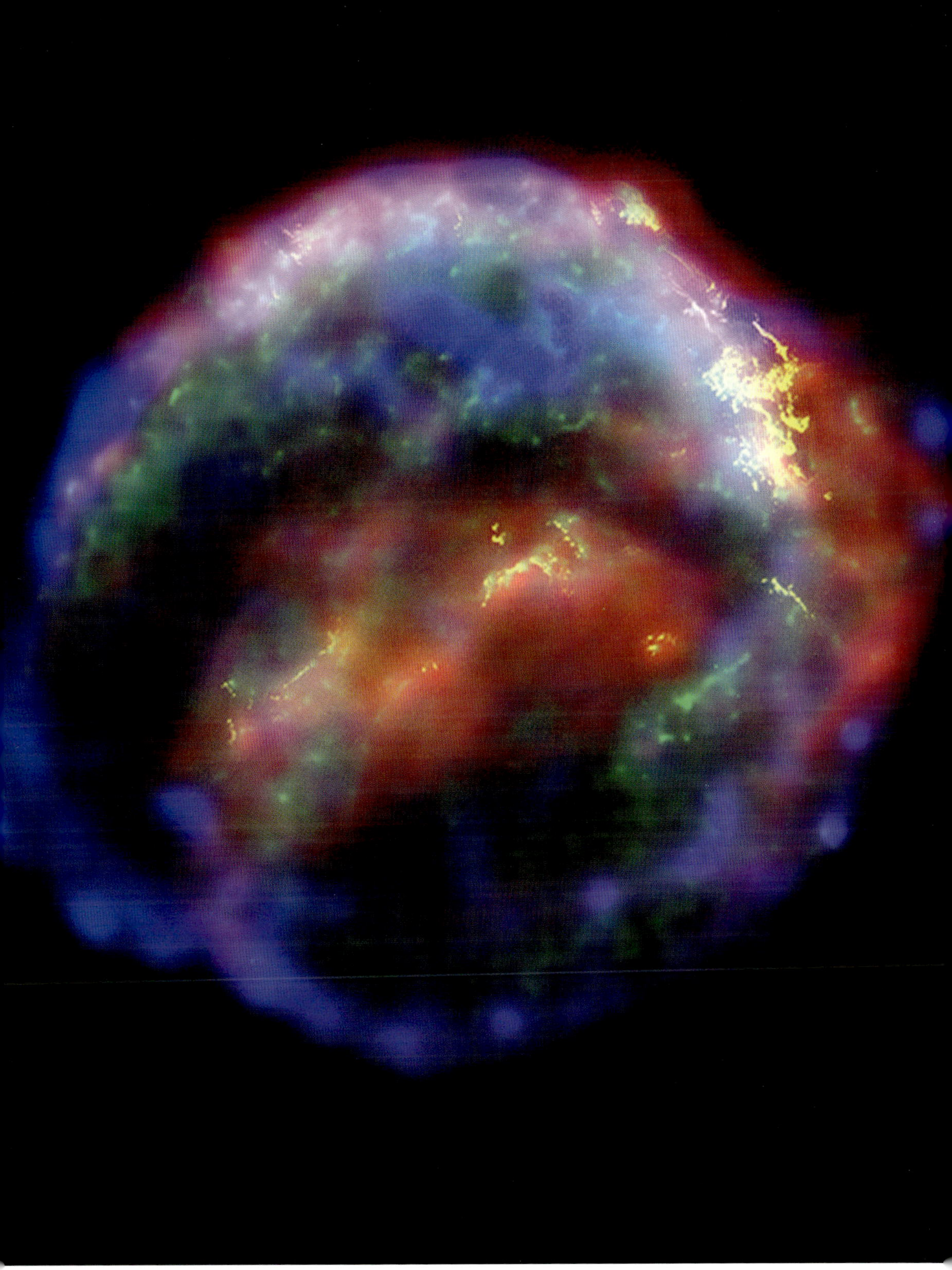

A false colour composite (CXO/HST/Spitzer Space Telescope) image of the supernova remnant nebula from SN 1604, likely to be host to a neutron star. (NASA/ESA/JHU/R. Sankrit & W.Blair)

FACT 19

The largest supercluster is home to 26 quadrillion stars

The Earth revolves around an average star, the Sun, which in turn orbits the centre of the Milky Way Galaxy. The Milky Way is a fairly average spiral galaxy with an estimated 100 to 200 billion stars and measuring 105,700 light years from side to side. That means if you stood at one end and turned on a torch, then travelling at 300,000 kilometres per second, it would take 105,700 years for the light to be seen at the other side of the Galaxy. Beyond our familiar cosmic backyard lies the vast expanse of the universe.

In my public lectures, I often derive a peculiar satisfaction from informing audiences that galaxies do not exist! As noted in Fact 1, they are not individual items like a cup or a car, a tree or a bus, but rather collections of celestial bodies bound together by gravity. In the same way, my home city of Norwich also does not truly exist. It is a collection of houses, cars, people, buildings, etc. Remove them and the city does not exist. Remove the contents of a galaxy and they too would not exist. Gravity makes them!

Gravity binds all galaxies together into giant clusters. Our own cluster is known as the 'Local Group' and consists of at least 80 galaxies. It does not stop there though; clusters are bound together by gravity into superclusters. Our Local Group cluster is a member of the Laniakea Supercluster. Spanning 250 million light years and encompassing an estimated 100,000 galaxies, Laniakea seems enormous but, compared to the Einasto Supercluster, it is a mere blip on the celestial landscape.

Named after Professor Jaan Einasto (b. 1929), who was a real trailblazer in supercluster research, this greatest of all superclusters is 3 billion light years away and spans 360 million light years. Think those numbers are big? The Einasto Supercluster is thought to contain 26 quadrillion (26,000,000,000,000,000) stars!

To understand the sheer scale of these immense structures, imagine a coin in the centre of a football pitch. The coin represents the Milky Way Galaxy and the pitch represents the expanse of the supercluster. An alternative way to envisage it is to imagine the Sun as a golf ball – then the combined mass of the supercluster would be about the same as Mount Everest! These things are among the most massive, giant structures in the universe.

The Einasto Supercluster was found after astronomers from the Tartu Observatory in Estonia surveyed 662 superclusters across the cosmos. They revealed many fascinating dynamics within the clusters. One particularly interesting observation was that galaxies in superclusters seemed to generally be receding slower from each other when compared to the general expansion of the universe. The reason for this slower velocity is the force of gravity which acts as a break on their movement, slowing the expansion of space in the local vicinity of the supercluster. This action is not sufficient however to halt the

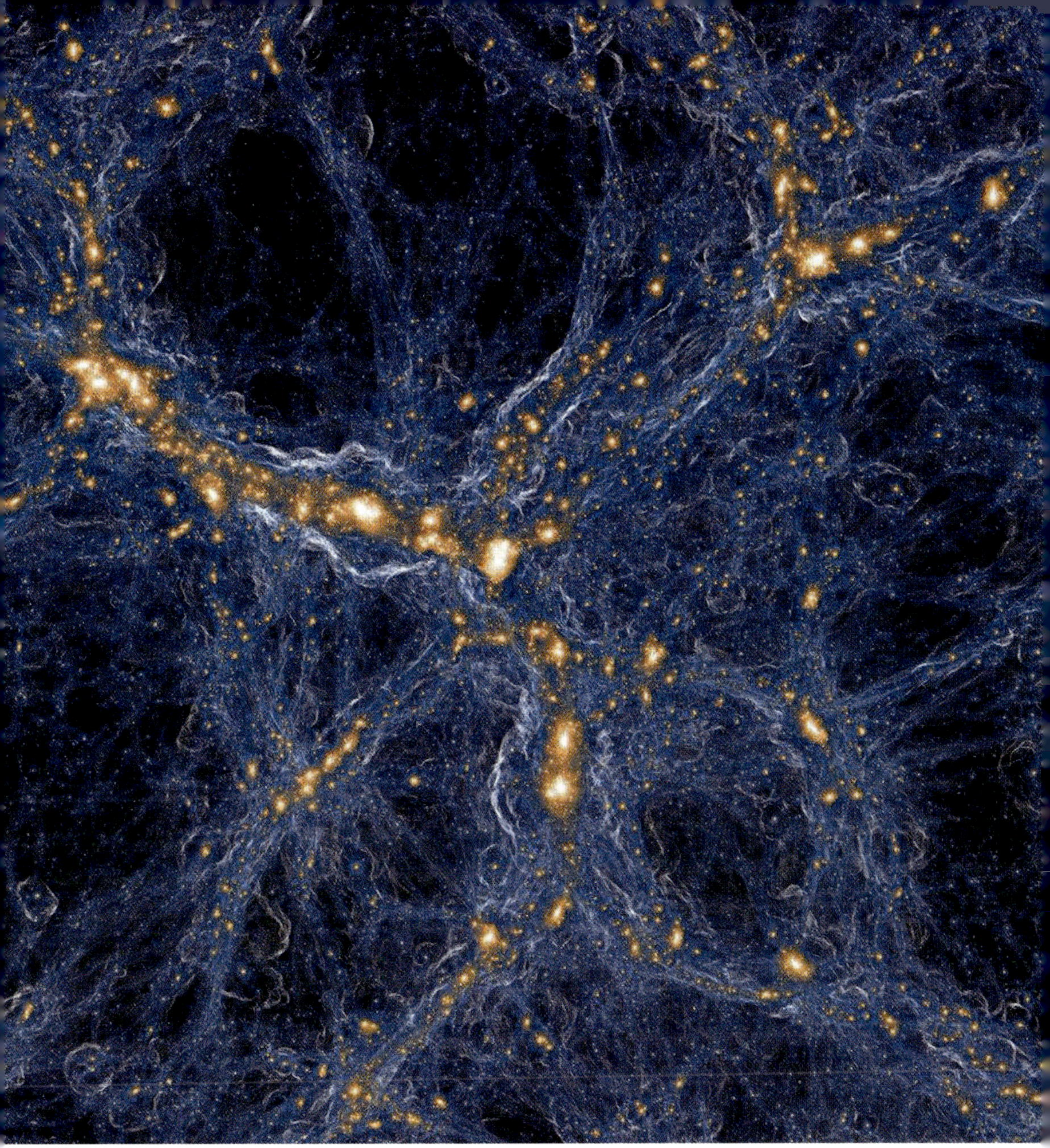

Visualisation of the Laniakea supercluster, which represents a collection of more than 100,000 estimated galaxies spanning a volume of over 100 million light years. (Tsaghkyan/Wikimedia Commons)

expansion and, over the fullness of time, the galaxies will drift further and further apart.

Another interesting observation was the relationship between the density and size of superclusters. The inverse square law seemed to describe this relationship nicely with increasing size bringing with it a decrease in density. The study of these superclusters helps us to not only understand the evolution of the structure of galaxies but also the large-scale evolution of the universe.

FACT 20

Barred spiral galaxies appeared early, just 2.7 billion years after the Big Bang

Anyone familiar with astronomy will know that galaxies come in a fairly limited range of shapes, typically spiral, elliptical, barred spiral and irregular. Barred spiral galaxies seem to be a feature of the modern universe, meaning that they may only have appeared relatively recently in galactic terms, but a study by astronomers using the James Webb Space Telescope (JWST) has found that the bar feature in some spiral galaxies seems to have appeared as early as 2.7 billion years after the Big Bang, suggesting galaxies evolved faster in the early universe than previously expected.

Our own Galaxy, the Milky Way, is a spiral galaxy with a central nucleus and spiral arms emanating out from the centre. As noted earlier in Fact 13, a great way to visualise this is the classic fried egg model – imagine two fried

NGC1300 is a great example of a barred spiral galaxy. (NASA/ESA/Hubble Heritage)

eggs stuck back to back and that's the broad shape of our Galaxy. The core is represented by the yoke of the eggs and the spiral arms emanate out as the whites of the eggs. Our solar system, with Sun, planets, comets and asteroids, lies about 25,000 light years from the centre.

If however you look at other galaxies in the sky, you will see a real mix, but generally they fall under the four main categories. Edwin Hubble (1889–1953) tried to bring some structure to the different shapes by developing his galaxy classification scheme to articulate not only the shapes but also the sub-categories within them. These sub-categories define, for example, how egg-shaped or spherical an elliptical galaxy is or how tightly wound the arms are in a spiral galaxy.

It has been known for some time that galaxies are not static, they move relative to each other and they evolve and change as stars are made and stars die. With most spiral galaxies, for example, as they age, they often develop a bar feature to replace the bulge of the core. The bar joins up the spiral arms and is believed to be only temporary, forming only when a build-up of gas creates a burst of star formation.

Understanding just how the bar feature forms is key to understanding the evolutionary process of the galaxy itself. All previous observations have showed that the appearance of the bar significantly reduces from the nearby universe to greater distances and, because light takes time to travel across the universe, observing more distant galaxies means looking further back in time.

Recent observations from the sensitive JWST reveal that some galaxies further away than expected do indeed have this bar feature. A team of astronomers visually searched through 368 galaxies that had appeared in previous Hubble Space Telescope surveys and counted those with a bar. They found that, as expected from the nearby galaxies, about 17.8% displayed a bar feature while this reduced to 13.8% at the greater distances.

The study revealed that the JWST's infrared sensitivity picked up twice as many barred spiral galaxies than Hubble's more blue-sensitive imaging platform. Working the maths, the team of astronomers worked out that barred spiral galaxies seem to have appeared during a much earlier epoch, around 11 billion years ago.

FACT 21

Black holes can kill galaxies

It is difficult to actually visualise a universe that is changing. Things tend to happen at snail's pace, albeit with the odd exception. Take the formation of galaxies growing in the early universe. Their immense gravitational field would suck in dust and gas from around them, creating vast collections of stars. In the very centre of these young galaxies, supermassive black holes would reside, turning the galaxies into powerful quasars.

Artist impression of the James Webb Space Telescope that captured data from wind flowing from black holes. (NASA)

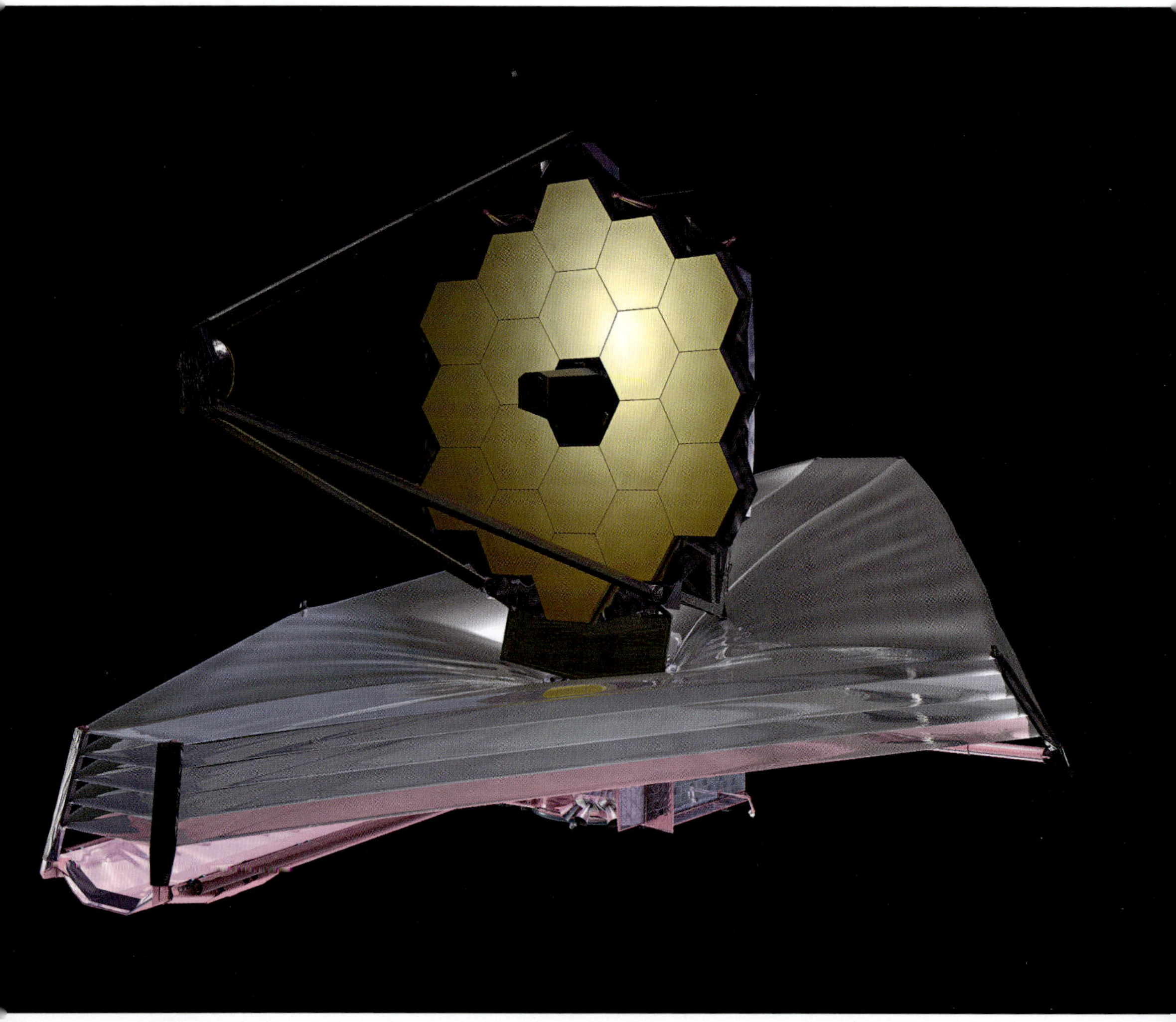

To remove the confusion and mystique around black holes: they are the corpses of massive stars. During the main part of the life of a star, fusion in the core generates incredible amounts of energy as new elements are synthesised. Along with new elements, heat and light, a powerful outflow of electrically charged particles rushes away and permeates the surrounding space. Here on Earth, charged particles reach us and on arrival we experience the glorious display of the Northern Lights.

As our Sun evolves, it will reach a stage when the pressure from fusion deep in the core overpowers gravity and it expands to such a great size that gravity is unable to hold on to the material and it drifts off into space. When stars far more massive than the Sun die, gravity overpowers the force from fusion and the core collapses. As the core collapses, all the rest of the stellar material has nothing to hold it up so it rushes down at eye-watering speeds onto the core, collapsing it even further. This leads to a rebound and the star explodes as a supernova. The remains of the core is a point source that is so incredibly dense that even light, travelling at 300,000 kilometres per second, is unable to escape. This is a black hole and it is believed that many galaxies have supermassive black holes at their core.

At varying stages of a star's life, the outflow of electrically charged particles drifts through the galaxy. But of all the 'wind' that flows outward, 90% of it is made of neutral gas. However, the James Webb Space Telescope (JWST) is now able to detect this gas where before it was invisible to us. The rest was charged, warmer and easy to detect. The powerful outflow of neutral gas comes from the supermassive black holes at the core of some galaxies at the edge of the universe.

A team of astronomers first identified that black holes were driving the neutral gas outflow by studying a distant galaxy over 10 billion light years away. There is a problem though – the outflow is removing gas faster than gas can coalesce to form new stars. The outflow is likely to have a very significant impact on the evolution of the galaxy.

With a lack of gas and dust, star formation will slow and eventually stop. Just like a forest that always has new trees growing to replace old, dying trees, so galaxies usually have star formations to replace dying stars. Ultimately, however, the forest, and a galaxy, will be unable to grow and develop and eventually become static and slowly die, with the final stars eventually blinking out.

FACT 22

We can measure the magnetic field of distant planets using radio telescopes

Life on Earth would not be possible without food, water, light, a breathable atmosphere and – surprisingly – a magnetic field. Without it, Earth and its inhabitants would be subjected to the harmful radiation from space, making life here impossible. If we find planets around other stars, or exoplanets with similar magnetospheres, then those worlds may well be habitable. The Square Kilometre Array (SKA) is an international project to build the world's largest radio telescope, with components in Australia and South Africa and its headquarters in the UK. When complete it should be able to detect such magnetospheres from radio emissions, giving us real insight into our exoplanet cousins.

The magnetic field of Earth is the result of the churning motion of liquid iron and nickel in the outer core. The resultant magnetic field has the properties of a giant magnet with a north pole and a south pole, and it extends from the core outward, enveloping the entire planet. The presence of this field stops harmful solar radiation and cosmic particles. Magnetic fields are not static though, and it is not uncommon for them to flip, as has happened to our own magnetic field.

Since we have been hunting exoplanets (and to date, over 5,000 have been discovered) it has become clear that there are a good number of super-sized gas giants. As our detection technology and methods improve, smaller, more Earth-like planets are starting to be discovered. It is therefore not unreasonable to think that, among them, there may well be alien planets with magnetic fields, making them, therefore, good candidates for habitable environments.

Understanding exoplanet magnetic fields is in its infancy. So far, we have only explored magnetic fields around the planets in our solar system. What we do know is that any planetary magnetic field emits radio signals due to the electron cyclotron maser instability (ECMI) mechanism. It sounds like something out of *Star Trek* or *Star Wars*, depending on your preference, but either way electromagnetic radiation is amplified by electrons that are trapped in the field. It is this amplified radiation that can be detected remotely *if* we have a radio telescope with the capability, which is where the SKA comes in. When complete, the SKA telescopes can be joined together electronically to operate as one big telescope covering a 'collecting area' of a square kilometre of 'radio sky' with higher sensitivity and resolution than ever before.

Using NASA's exoplanet archive data, astronomers have calculated the strength of radio signals from 80 confirmed planets. They took the planets' radius, mass and orbital distance from the host star, along with the star's mass, radius and distance from us to estimate the signal from the

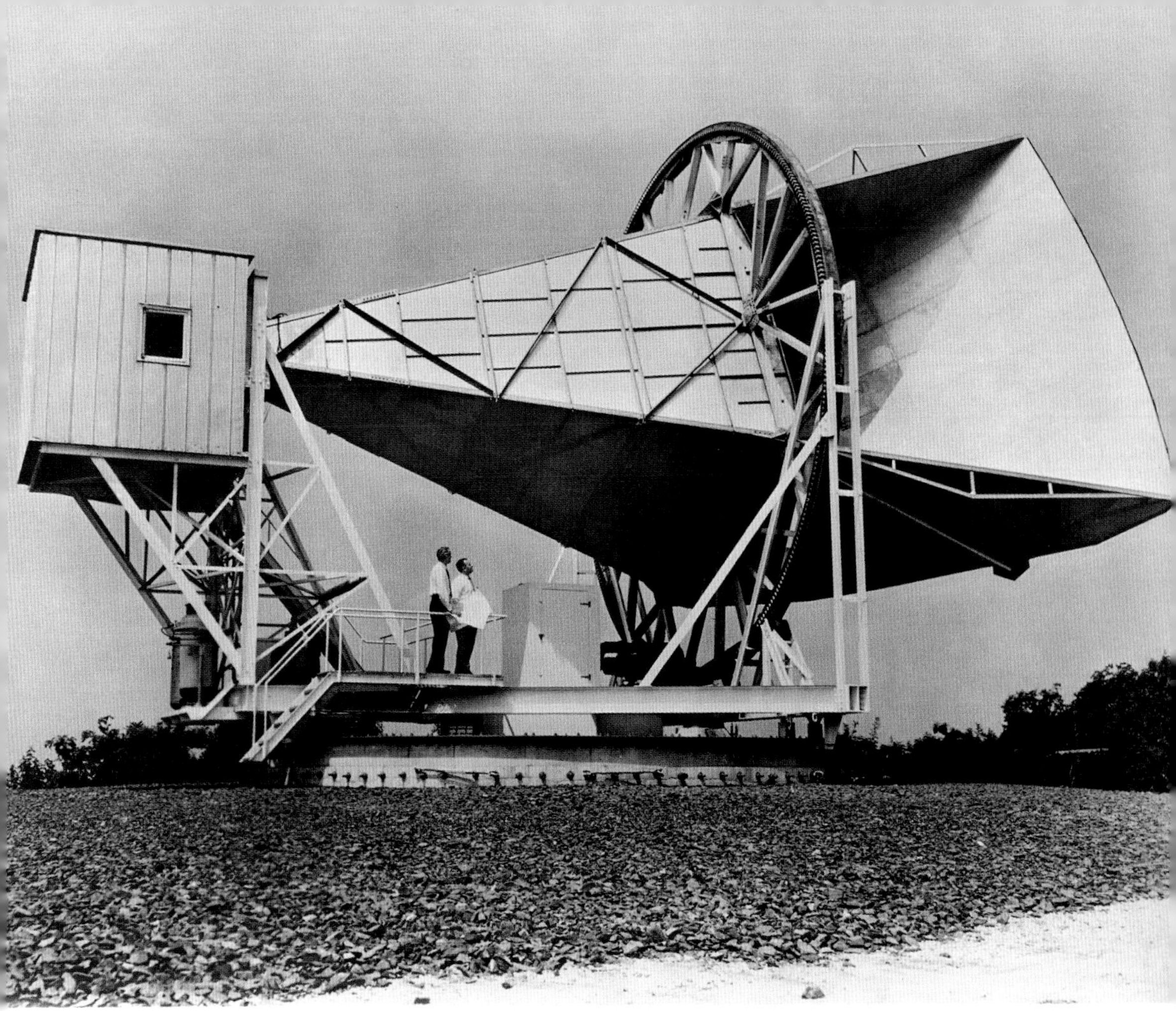

The Holmdel horn antenna at Bell Telephone Laboratories, New Jersey, where in 1964 Robert Wilson and Arno Penzias accidentally discovered cosmic microwave background radiation — the echo of the Big Bang — earning them the 1978 Nobel Prize in Physics. (NASA)

magnetosphere. The results were promising and confirm that exoplanets Qatar-4 b, TOI-1278 b and WASP-173 A b would indeed emit radio signals from their magnetosphere that the SKA could detect. Unfortunately we will have to wait until 2028 when the SKA is operational but already it seems researchers are lining up to use it and this piece of research in particular looks set not only to herald a greater understanding of exoplanets but also of the possibility of life in the universe.

FACT 23

We can see the echoes from the Big Bang

Ask any astronomer where the universe came from and you will more than likely get the answer that everything – the Sun, the Moon, the planets and galaxies – all come from an event known as the Big Bang. You might get a slightly less certain answer if you ask what came before that, and to be frank, there is no consensus. Certainly there are theories, many of which require you to disconnect your common sense, but for now, a solid theory really does elude us. We can be far more certain about the Big Bang however.

It may be difficult to get your head around the concept and even struggle to believe that the universe does indeed have a beginning. After all, when you look at the night sky, it seems pretty much never-changing. It is not too surprising though, because most things in the universe happen at a bit of a

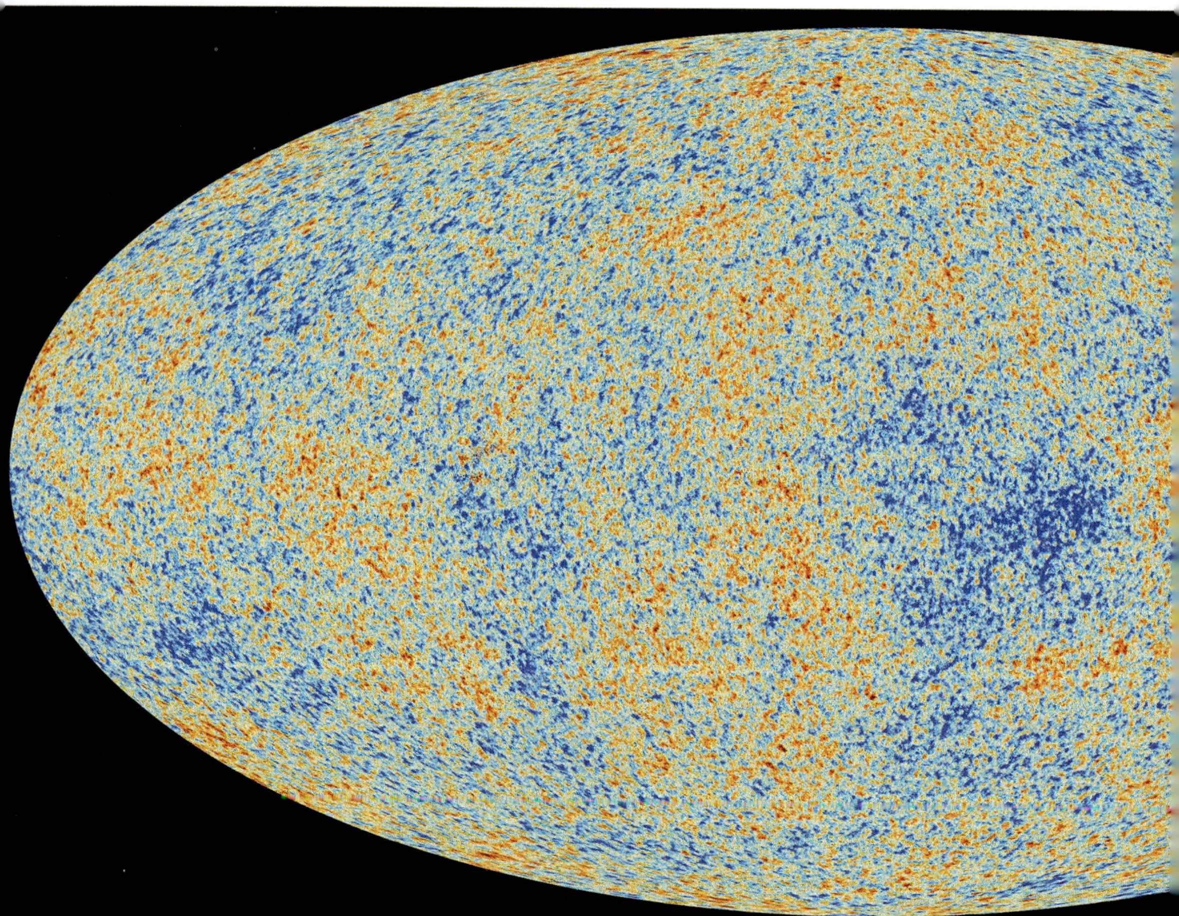

snail's pace. However, just like a walk through a forest it would be possible, with a little lateral thinking, to work out how the forest evolves and changes by studying trees in all their different stages of life. There would be young saplings, old trees and even dead trees, and from that you could, for example, infer the life cycle of a tree without watching it grow and die. By observing the universe and taking on board all the evidence, we can work out what has happened and what might happen in the future.

One observation that led to the Big Bang theory was the motion of the galaxies. Using an instrument known as a spectroscope it is possible to split the light from distant galaxies into its component parts. When we do that, it is possible to see the signature lines of different gasses, but they do not appear in their usual places, they are *shifted* toward the red end of the spectrum. This *red shift* is the result of the movement of galaxies and, with the few exceptions of nearby galaxies, *every* other galaxy in the universe is moving away from us or, more accurately, every galaxy is moving away from every *other* galaxy. You can visualise this by imagining dots on the surface of a balloon. As the balloon is blown up, every dot will move away from every other dot, none are at the centre. In the same way, as space expands, all the galaxies are carried away from each other. Tracking backwards in time we can calculate that 13.7 billion years ago they must have all been occupying the same region of space, the moment of the Big Bang.

The movement of the galaxies is not the only evidence of the Big Bang. If we study the sky in microwaves, we can detect a glow across the entire sky. After the Big Bang there was a brief period when the universe was opaque and light could not travel through it. Eventually it could travel freely across space and, as space continued to expand, the visible light waves became stretched out to microwave radiation wavelengths. It is this radiation that microwave telescopes can detect, echoing out across the universe.

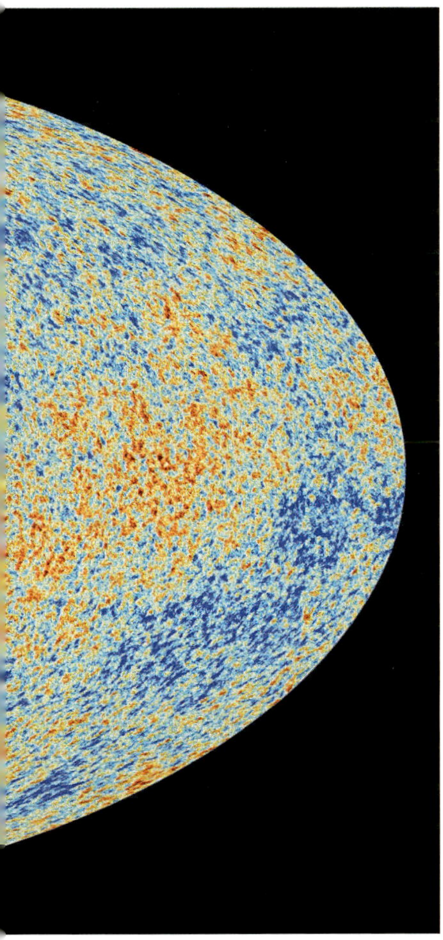

Temperature map of the cosmic microwave background measured by the Planck spacecraft. (ESA)

FACT 24

Some astronomers think a signal from ET was received in 1977

Among the many unanswered questions in science, maybe even society, is the question of whether we are alone in the universe. For decades now, we have been in search of signals from alien intelligences. The challenge is a million times worse than looking for a needle in a haystack though. First is the challenge of pointing in the right direction at the right time and second is tuning in at the right frequency. Get those right though and, assuming someone is sending a signal too, you have the answer that has been plaguing society for centuries.

The hunt for 'ET' (extraterrestrial life) is not a particularly new thing. Even futurist and inventor Nikola Tesla (1856–1943) suggested that a wireless electrical transmission system could be used to contact aliens on Mars. He even thought one of his experiments had picked up a signal! Today, we tend to favour radio signals rather than any other type of radiation since they do not get blocked by dust clouds which permeate the Galaxy. There are one or two optical surveys looking out for alien signals that may be sent by giant lasers but it is radio signals where the real focus lies.

To that end, great big radio telescopes across the planet relentlessly scour the skies in the hope of picking up a signal. One such telescope dedicated to this search was based at Ohio State University. Nicknamed 'Big Ear', the telescope had been conducting a survey of extragalactic radio sources and

The WOW! signal represented as 6EQUJ5 on the original computer printout.
(Ohio State University Radio Observatory)

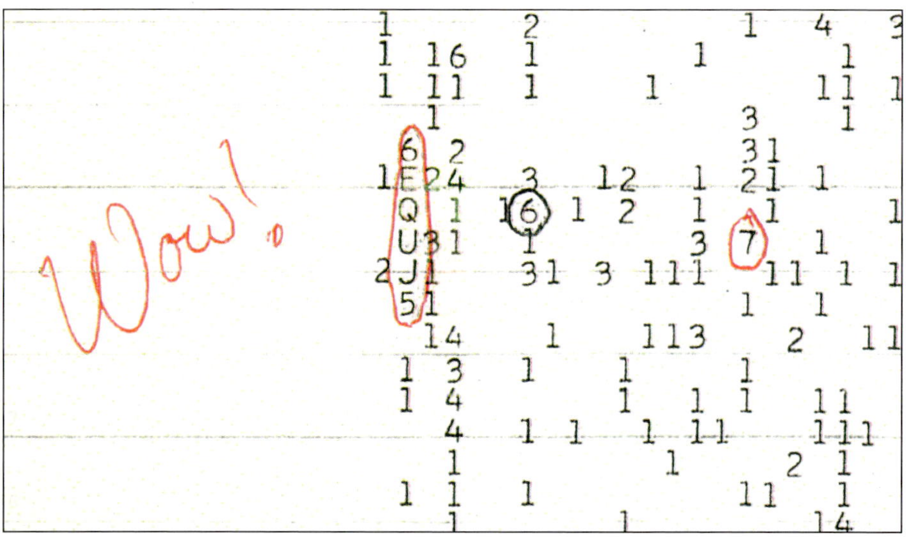

in 1973 was assigned to the search for ET. The telescope was a Kraus-type instrument with a design based on that by Dr John Kraus. It was a transit instrument that had a horizontal collector which reflected radio waves towards a receiver. Signals were collected as objects tracked over the observatory.

Typically, radio telescopes will tune in to a frequency of 1420 megahertz, as that frequency is emitted by hydrogen (the most abundant element in the universe) so any technologically advanced civilisation would be familiar with it. It was around this frequency that Big Ear was tuned on 15 August 1977. The telescope was recording data when Sagittarius was overhead and a few days later astronomer Jerry Ehman (b. 1939) was checking through the data. The data presented was nothing more than strings and strings of numbers where the higher values recorded a stronger signal. Any signal with a value of 10 or higher was recorded as a letter: A being the lowest where A = 10, B = 11 and so on, up. The majority of the data recorded as 1s and 2s, a background signal, but for a period of 72 seconds, the signal suddenly leapt to 6EQUJ5, recording a peak signal about 30 times higher than the background noise.

Ehman was so taken by the signal that he circled it and scribbled 'WOW!' in the margin, meaning an exclamation of surprise. Was it a signal from an alien intelligence? It certainly could have been but there may have been other explanations and, given that no repeat of the signal has been observed since, we may never know.

FACT 25

The Milky Way might have too many satellite galaxies

The Large and Small Magellanic Clouds are well known satellite galaxies of the Milky Way but you might not realise that there are more! The Milky Way is surrounded by at least 61 within 1.4 million light years (for context, the Andromeda Galaxy, the nearest major galaxy, is 2.5 million light years away), but there are likely to be even more. Sky surveys have revealed nine more satellite galaxies in only 3% of the sky!

Data from *Gaia* (a satellite collecting accurate position information of astronomical objects) suggests that most of the satellite galaxies orbiting our own Milky Way are newcomers. Even the Large and Small Magellanic Clouds are now known to be newcomers even though they have been in our sky for most of recorded history. The relative motion of the two shows they are not currently in orbit around us but may, some time in the distant future, leave our cosmic neighbourhood. They may of course stay, but whether they or any others will fall into an orbit around our Galaxy is as yet unknown, largely because we do not have an accurate measure for the mass of the Milky Way.

Central to our understanding of local galaxy distribution is a drive to understand *dark matter* distribution across the cosmos. The concept of the universe being dominated by cold, dark matter nicely describes its large-scale structure. It struggles however, to describe the structure in the local universe predicting hundreds of satellite galaxies to the Milky Way. Observations have only revealed a handful of satellite galaxies contradicting the model in a quandary known as the 'missing satellites problem'.

Previous data obtained before 2018, of an area of sky covering a massive 676 square degrees, revealed three candidate satellite galaxies: Vir I (in Virgo), Cet III (in Cetus) and Boo IV (in Bootes). Data released over the three years that followed covering 1,140 square degrees revealed two additional candidates: Sext II (in Sextant) and Vir III (in Virgo).

These studies are of course based upon statistically small numbers and small areas of the sky, and several assumptions had been made based on an expected distribution of satellite galaxies. It does not alter the fact, however, that to date there seem to be far too many satellite galaxies than the dark matter model predicts. Like many aspects of science, we often have to readdress theory to meet observation – it is all part of the scientific journey.

The Large Magellanic Cloud. (ESO/VMC)

FACT 26

Fish could turn regolith into fertile soil on Mars

Mars is not typically classed as in deep space but this is a fascinating discovery that might serve us well on exoplanets too. Here is the problem: we travel to Mars to try and colonise it, but how do we feed ourselves? Sure, we can take a truckload of food with us but for the return trip that is a lot. If we plan to colonise the red planet, or indeed any other planets in deep space, we need a better, more sustainable solution. We have to grow or somehow create food while we are there.

In the next few decades we may well see the colonisation of Mars. It is 54.6 million kilometres away so, even on board a rocket, it takes about seven months to get there! Future colonists could simply have supply ships drop all they need, but that becomes ridiculously expensive to sustain and frankly, is not sustainable. The people that colonise Mars will just have to find some way to grow what they need.

If you have watched *The Martian* movie with Matt Damon you will know how unforgiving the Martian environment is. OK, the film was a little out on scientific accuracy in places but it certainly showed how inhospitable it really is there. Matt managed to cultivate a decent crop of potatoes in Martian regolith (the top layer of Martian soil), fertilised with human faeces. This may not be quite so practical in real life and there may be alternative, less smelly, and dangerous, alternatives.

Taking the assumption that colonists will have to grow fresh produce locally, a solution is needed. On first glance, it may seem not too great an idea after all – the atmosphere on Mars is toxic, with 95% carbon dioxide (compared to just 0.04% on Earth). There is a similar length of day on Mars but being able to grow crops will require longer periods of lighting. It is possible that water may be collected from the ice which forms on and in the Martian rocks.

Putting the atmosphere to one side, since it would require some form of biosphere to enable the inhabitants to live, attention has turned to fish and how they could help impart nutrients into the Martian regolith. To test the idea an aquaponic system has been used with fish in tanks to generate the nutrient-rich liquid, with the liquid being tested on simulated Martian soil.

The results were very promising. An aquaponic system not only facilitates growing plants within the system itself but the nutrient-rich water also performed as an excellent fertiliser. This took the organically deficient regolith and turned it into something akin to useable soil. Using a fish known as tilapia (*Oreochromis niloticus*), potatoes, tomatoes, beans, carrots and much more have been grown in tests. The plants were grown and indeed thrived in a tent that simulated Mars in every way possible here on Earth.

Buzz Aldrin's boot print in the lunar regolith. (NASA)

FACT 27

It is possible that heart attacks may be more common in space

Astronauts are considered by many to be an elite bunch of people – healthy, fit and capable in many disciplines. When they travel into space they can face health issues related to weightlessness from reduction in bone density, to problems with their eyesight. These are people at the peak of physical fitness but what will happen to the rest of us when space tourism really kicks off?

Space travel and artificial intelligence (AI) are two fabulously interesting topics. Combine them and you have a fascinating story that has gone some way to address this very concern. Using AI and mathematical models, scientists have explored human physiology and the impact of space exploration on a couple of twins, one with a heart condition.

As space travel becomes more available to the mass population we will start to see a shift in the demographic of space travellers to older, more wealthy individuals, but they are more likely to have health problems. We will eventually see people with a whole multitude of conditions wanting to holiday in space, but what are the likely impacts?

The real concern is the experience of *microgravity* which causes a redistribution of fluids around the body and can cause conditions like 'puffy face, bird leg syndrome'. The name aptly describes the effect: the face swells up and the legs thin. It results in an increase in venous pressure in the upper body which is fine for healthy people but heart failure sufferers are at a much higher risk. Given that there are over 100 million people around the world who suffer from heart failure it is essential this is explored.

How big is the problem though? Looking at the wide spectrum of heart failure, conditions can be grouped into two categories: a weak heart that cannot pump effectively and a heart that cannot relax and fill properly. All possible conditions (and there are of course more than two) need to be studied to come up with specific ways to treat and mitigate the risk during space travel.

It is difficult to collect real data in space to understand this problem, so AI comes to the rescue to model and simulate the effects. The study with the twins showed that a microgravity environment leads to an increase in cardiac output (the quantity of blood pumped by the heart in a given period of time). This is not normally a problem but with heart failure patients it is accompanied by a rise in pressure in the left atrial region of the heart to dangerous levels. If left unchecked, it can lead to a condition where fluid accumulates in the lungs, known as a pulmonary oedema, making it difficult to breathe.

With the increase in corporate interest in space travel, space tourism is slowly becoming a reality. People can already pay for trips into space, but as costs come down the number of people heading out into space will increase. Eventually, trips into space will be as common as trips to other countries. It is imperative we understand the impact on our health and what we can do to make space as widely accessible as possible without putting our health at risk.

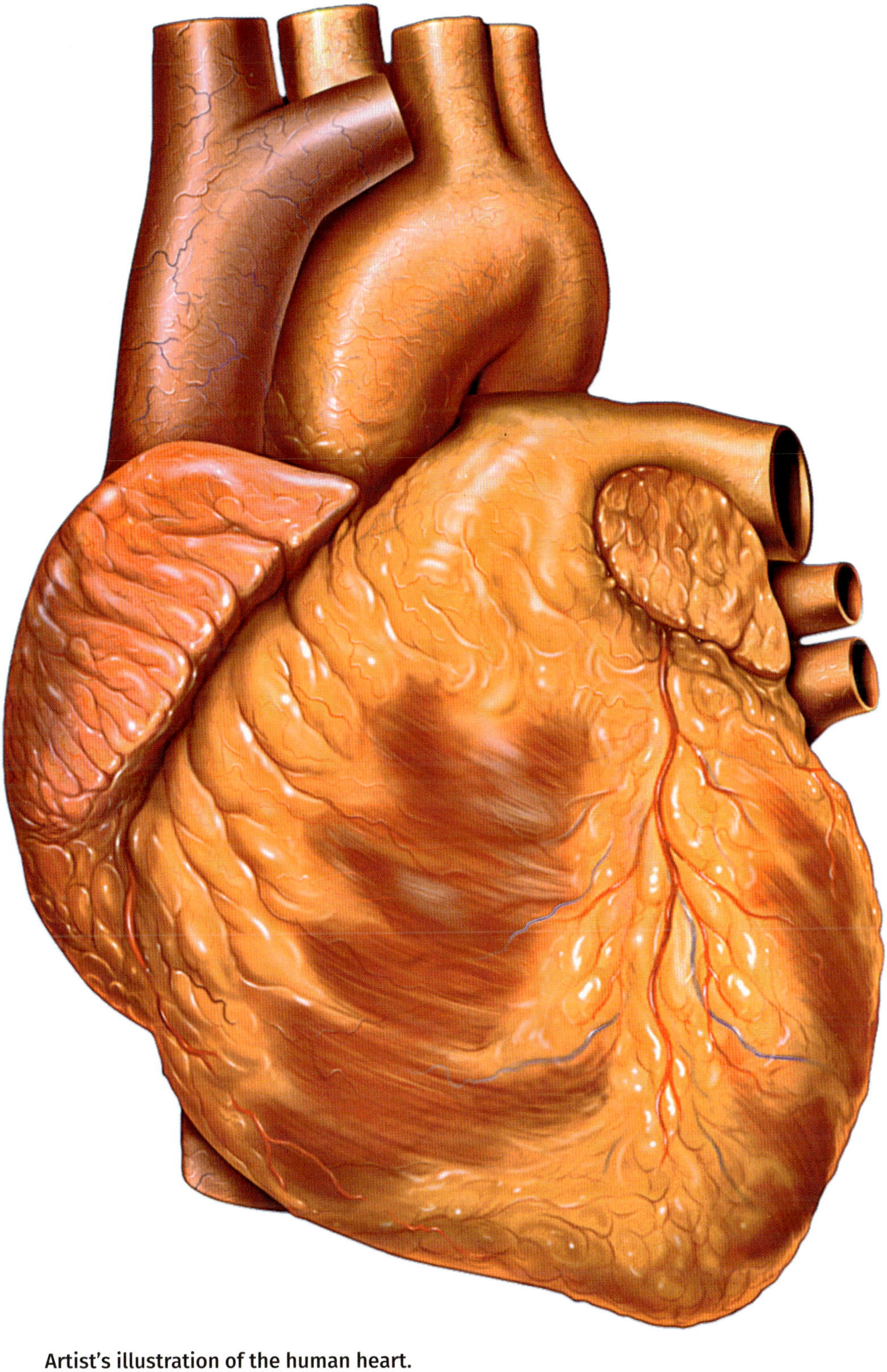

Artist's illustration of the human heart.

FACT 28

Merging galaxies make for explosive star formation

NGC4449 is a stunning irregular galaxy that shows high levels of star formation. This galaxy is categorised as a 'Magellanic type' galaxy due to its similarities with the Magellanic Clouds, although it is smaller in size.

All stars, even the Sun, are born from vast clouds of gas and dust, and when they die, their remnants are recycled back into the galaxy, providing fuel for new stars. When the universe began there was only hydrogen and a tiny amount of helium present but the fusion process inside stars creates new, heavier elements. This includes every atom inside you and me, the planets and even the computer screen or paper you are reading this on. In a relatively nearby part of the universe, 13 million light years away in the constellation Canes Venatici, this cycle is beginning again at an extraordinary pace.

NGC4449 (the four thousandth, four hundredth and forty ninth object in the *New General Catalogue*, NGC) is a member of the M94 group of galaxies and is surrounded by a halo of dwarf galaxies, two of which are currently merging with it. The Gemini Observatory has recently captured incredible images of NGC4449, showcasing the star formations or 'birth processes' occurring within.

Deep within NGC4449, stars have been forming for billions of years. It is producing stars now at a significantly accelerated rate leading to its reclassification as a *starburst galaxy*. Starbursts are usually restricted to central regions but NGC4449's is much more widespread. The majority of hot young stars are now found in the galaxy's nucleus and surrounding regions.

Such levels of star formation are reminiscent of the star formation in galaxies in the early universe, although the driving force then was galaxy mergers and accretions. It is therefore thought that the burst of activity in NGC4449 is the result of a galactic merger or interaction with a neighbour.

Indeed, one neighbouring galaxy displays a faint stream of stars extending to the side of NGC4449, showing it is currently in the process of merging. The string (and satellite galaxy) are barely visible visually due to their low visual brightness but can be detected due to their interaction with NGC4449. Another object that suggests previous interactions is a huge globular cluster. It seems to be embedded within the outer halo of NGC4449 and is now believed to be a nucleus from a previous satellite galaxy that has merged with its companion. It is thought that the rate of star formation in NGC4449 is likely only to last for another billion years or so until the gas has been used up and the shockwaves subside.

Image of the starburst galaxy NGC4449. (NASA/ESA)

FACT 29

Many planets have the raw ingredients for life

Exoplanets are planets that orbit stars outside our solar system. They have been identified in the thousands since their first discovery in 1992, currently totalling 5,288. They vary widely in size, composition and orbit, ranging from gas giants like Jupiter to rocky, potentially habitable planets similar to Earth. Advanced telescopes and detection methods like the transit and radial velocity techniques have enabled the discovery of Earth-sized exoplanets. Their study not only enhances our understanding of planetary formation and evolution but also the search for extraterrestrial life.

That search for extraterrestrial life is no easy feat. Looking for aliens or at least environments where extraterrestrial life could one day evolve means knowing what to look *for*. To start with we can assume life has some

Artist's impression of the PLATO spacecraft. (ESA/ATG)

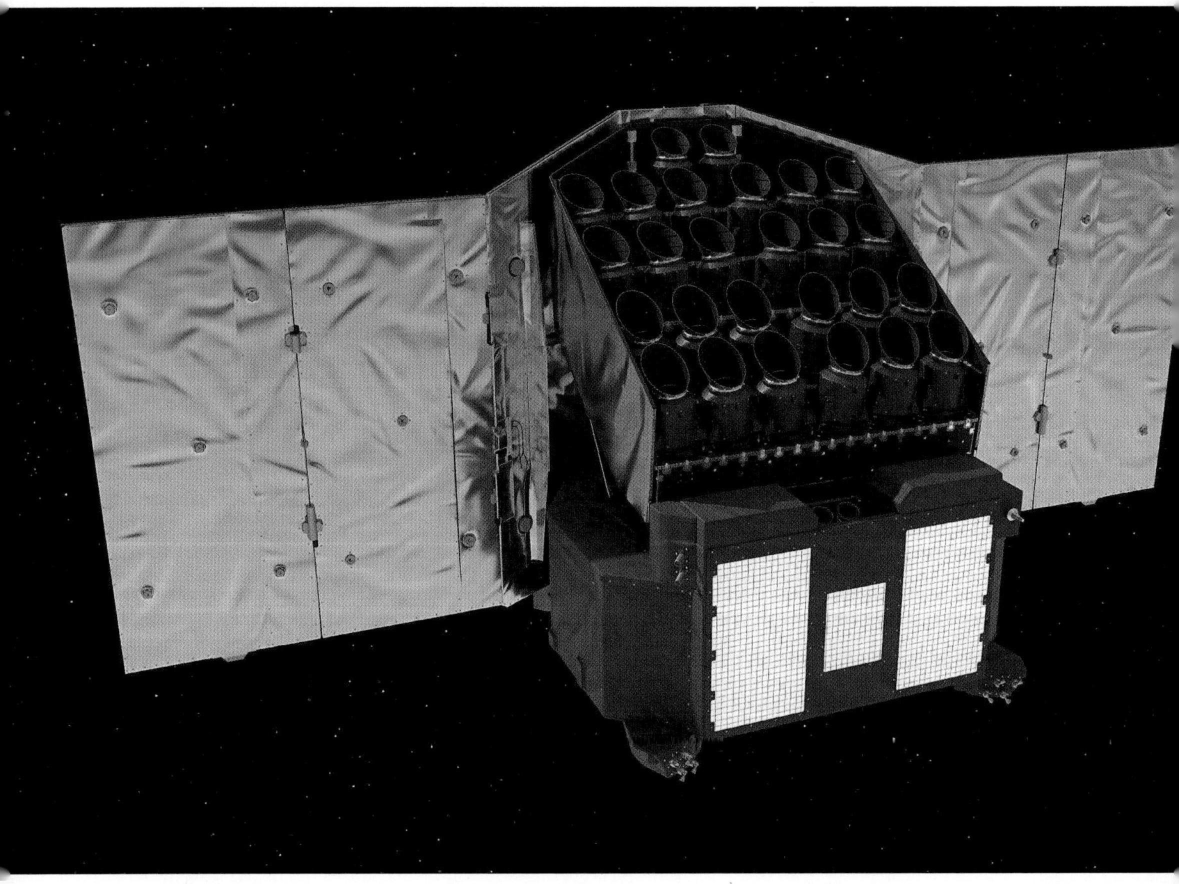

basic requirements or building blocks (CHNOPS: carbon, hydrogen, nitrogen, oxygen, phosphorous and sulphur) and a solvent to life's reactions (generally, liquid water). We have to assume that these or similar requirements might be universal in the cosmos. There is of course a chance of life based on a completely different set of needs but if we are going to start somewhere then we may as well start looking for life like that found on Earth, otherwise well, who knows what to look for!

Finding planets that already have, or have the ingredients for, intelligent life is a real challenge. It is exciting that new telescopes and spacecraft are in development that will start to identify candidate planets. Undertaking these observations, though, will take significant amounts of telescope time so we need to find some way to prioritise which ones to look at first.

Life on Earth can gain energy from a wide range of different thermodynamic disequilibria. A great example is life that thrives at the bottom of the ocean, taking energy and nutrients from thermal vents. More widely, life relies upon chemical reactions where an electron is lost or gained, changing its oxidation state. This is known as *redox disequilibrium*. Each reaction requires special proteins called oxidoreductases. The process requires metals as catalysts and without them it is unable to progress.

The distribution of these metals (which are more accurately known as *transition metals*) in the universe varies significantly over time and space. Despite this wide-ranging distribution across the cosmos, the role of these metals in enabling life has been largely overlooked in identifying astrobiological targets. A paper published in 2025 by Giovanni Covone and Donato Giovannelli proposed that the presence of certain elements is essential for habitability and should be prioritised as a primary factor when selecting exoplanetary targets in the search for life.

High resolution spectroscopy of rocky planets identified by missions like ESA's PLATO mission will focus on hunting for CHNOPS elements in stars. This data, along with the exoplanet parameters, will help hone the search. Identifying promising candidates will then enable follow-up observations as telescope and observatory technology becomes available over the coming decades.

FACT 30

There is a star that measures 11.4 kilometres across

The death of a massive stars leads to one of a number of different objects, dependent on the mass of the star. Two of them, though, are closely related: the neutron star and the pulsar. Both are formed during the core collapse and supernova explosion that marks the death of a supermassive star. All of the components of the atoms are squashed together, removing all the space between the neutrons to form one *massive* neutron.

Pulsars are rotating neutron stars with strong magnetic fields that emit beams of electromagnetic radiation from their magnetic poles. These beams become visible from Earth when aligned with our line of sight, creating a pulsating effect, hence the term 'pulsar'.

One of the nearest pulsars, PSR J0437-4725 lies at 510 light years in the constellation Pictor. It rotates 174 times per second which means it rotates once in just 5.75 milliseconds. Perhaps more mind-blowing than its rotational velocity is its size. Imagine 1.4 times the mass of the Sun squashed up into a ball just 11.4 kilometres across; the Sun is 1.39 million kilometres across by comparison!

This astonishing result of the discovery of the pulsar's diminutive size was the fruits of work by a team of astronomers at the University of Amsterdam. They used data from the NICER X-ray telescope on the International Space Station, combining it with a method called *pulse profile modelling* to get the result. The data was fed into Snellius, the Dutch national supercomputer, and complex statistical models were created. This allowed the astronomers to calculate the star's radius.

Not only were they able to work out its dimensions but also to map out the temperature distribution of the magnetic poles. They were able to show that the two hot magnetic poles were not directly opposite each other!

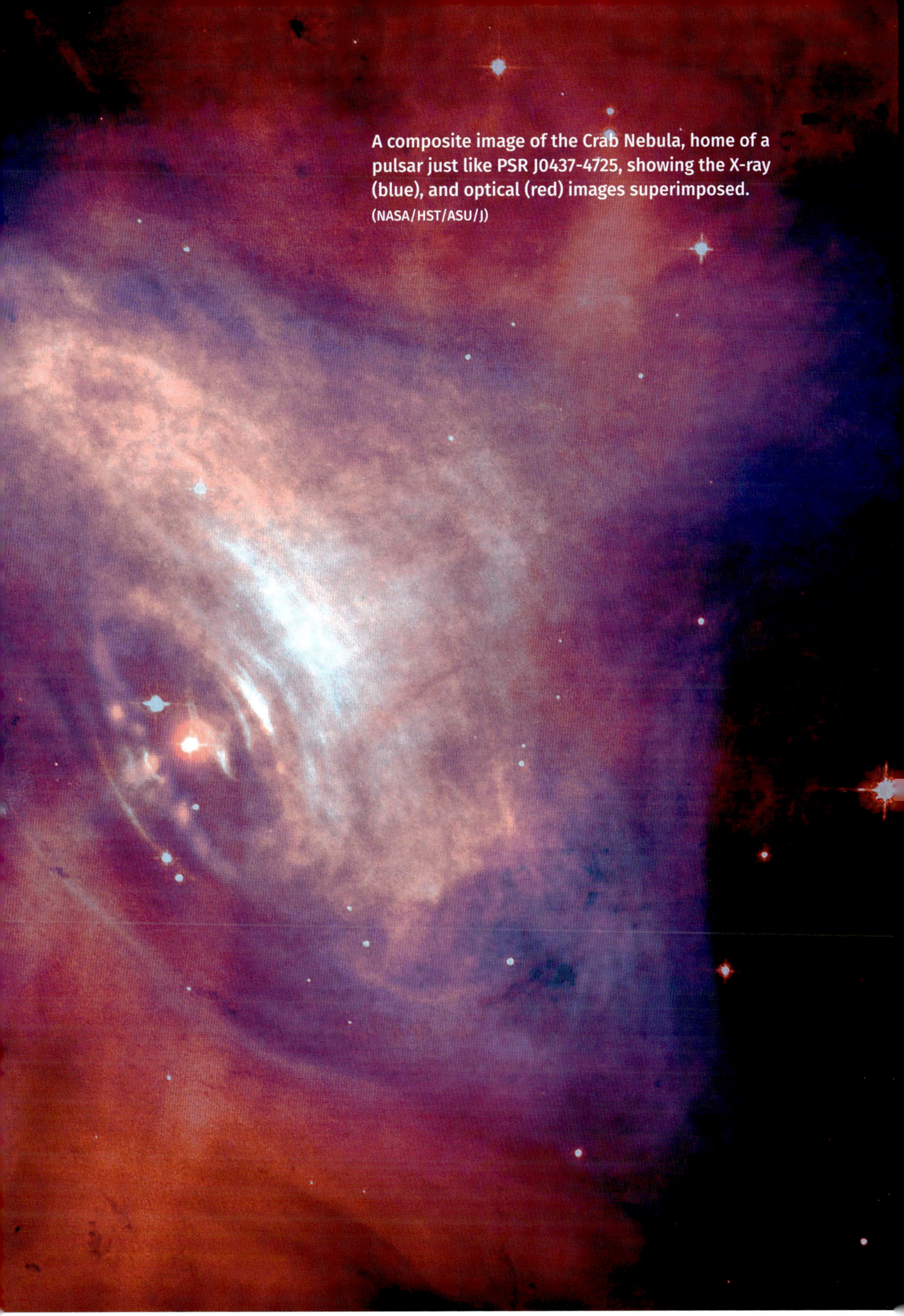

A composite image of the Crab Nebula, home of a pulsar just like PSR J0437-4725, showing the X-ray (blue), and optical (red) images superimposed.
(NASA/HST/ASU/J)

FACT 31

Stars' dwarf galaxies help us to map dark matter

Dark matter is one of the most mysterious components of the universe. As its name suggests, it is dark, meaning it does not emit, absorb, or reflect light – making it extremely difficult to study. Although it is invisible, its presence is inferred through its gravitational effects on visible matter. Scientists estimate that dark matter makes up approximately 27% of all the mass and energy in the universe, yet it interacts only gravitationally, remaining undetectable through traditional means.

The concept of dark matter first, ahem, came to light as an explanation for discrepancies in the rotation curves of galaxies – where stars at the outer edges of a galaxy orbit at unexpectedly high speeds. These observations suggest the existence of an unseen mass exerting gravitational influence. Despite decades of research, the exact nature of dark matter remains elusive. However, understanding dark matter is crucial for understanding both the composition and evolution of the universe.

Over recent decades, dark matter has been described as the invisible 'glue' that holds galaxies together. While galaxies contain significant amounts of visible matter, the distribution of dark matter within them provides key insights into its nature and role in galactic evolution. Computer simulations suggest that dark matter should be densely concentrated at the centre of galaxies, forming what is known as a *density cusp*. However, observations indicate that dark matter is more uniformly spread throughout galaxies, contradicting these predictions. Resolving this discrepancy is essential for refining our understanding of dark matter and the fundamental forces that shape the cosmos.

One of the primary ways scientists study dark matter within galaxies is by analysing the motion of stars. Since dark matter exerts a gravitational pull on stars, their movement provides indirect evidence of its distribution. Astronomers often rely on the Doppler effect to measure the speed of objects in space, observing changes in the wavelength of light as stars move towards or away from Earth. In addition to this line of sight movement, stars also move across the sky – a motion known as *proper motion*. By combining proper motion data with Doppler shift measurements, scientists can construct precise three-dimensional maps of stellar motion, which in turn help reveal the underlying distribution of dark matter.

NASA's Hubble Space Telescope has been instrumental in advancing our understanding of dark matter. Astronomers have used Hubble to study the Draco dwarf galaxy, a small galaxy located approximately 250,000 light years from Earth. Dwarf galaxies like Draco are particularly valuable for dark matter research because they contain relatively little visible matter, making the effects of dark matter more pronounced. Researchers analysed observational data over 18 years, from 2004 to 2022, to map the precise motions of stars within Draco. This long-term study has provided the most accurate measurements yet of how stars move in a dark-matter dominated galaxy.

The techniques developed for studying Draco's stars can be applied to other galaxies as well. Similar analyses are already underway for the Sculptor dwarf galaxy and the Ursa Minor dwarf galaxy. By refining these methods, astronomers hope to unlock more secrets of dark matter and its role in shaping the universe.

Astronomers ahve studied the movement of stars in the dwarf galaxy Draco (NASA/ESA)

FACT 32

We have a way to see whether the universe has been designed to support life

The *anthropic principle* (AP) suggests that the fundamental parameters of the universe, such as the strength of fundamental forces (gravity, electromagnetism, the strong and weak nuclear forces) appear to be finely tuned to support life. Whether this is true, or even a scientifically valid question, has been widely debated. Some argue that the AP is a crucial consideration in cosmology, while others dismiss it as a philosophical curiosity rather than a scientific principle.

The concept of the AP was first introduced by physicist Brandon Carter (b. 1942) in 1973 during a conference marking the 500th anniversary of the birth of Nicolaus Copernicus (1473–1534). Carter's proposal aimed to explain the apparent 'fine tuning' of universal constants that make life possible. The central idea is that if these parameters were even slightly different, life as we know it might never have evolved. There are two versions of the AP: the weak anthropic principle (WAP) and the strong anthropic principle (SAP). The WAP states that we observe a universe compatible with our existence simply because, if it were not, we would not be here to observe it! Think about that for a bit ... the SAP, on the other hand, makes a bolder claim – it asserts that the universe must have physical laws and parameters that make life possible.

Is the AP a scientific theory or merely a philosophical argument? A scientific theory should not only explain observations but also make testable predictions. Until recently, it was widely believed that the AP was beyond testing. However,

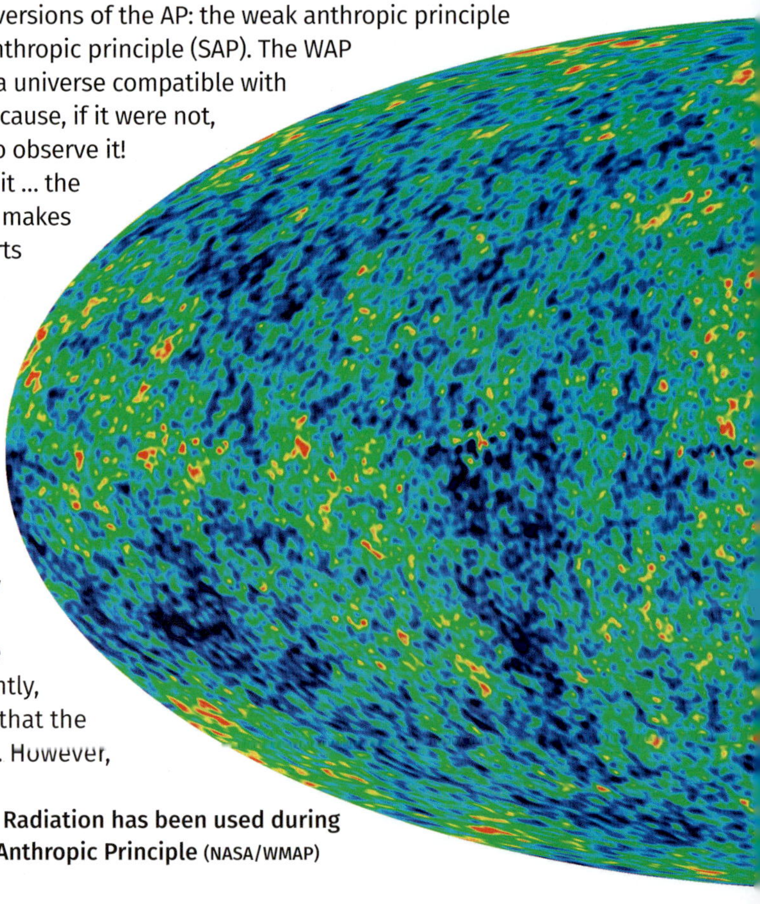

The Cosmic Background Radiation has been used during previous studies of the Anthropic Principle (NASA/WMAP)

at its core, the AP suggests that if the universe were to develop in a way that allows carbon-based life to evolve, it must have started with an extremely precise set of physical parameters. Key constants such as the gravitational constant, Planck's constant (which governs how energy must be in certain discrete amounts) and the elementary charge of an electron are examples of these parameters. If any of these values had been even slightly different at the beginning of time, the universe would have evolved in a dramatically different way – potentially one that could not support life.

Scientists have identified these initial parameters and used them to construct models of how the universe should have evolved, if the AP holds true. By comparing the predictions from these models with the actual structure of the observable universe, researchers can assess whether the AP provides a valid explanation of reality. Several predictions arising from the AP could be used as valid tests including aspects of cosmic inflation and the effects of dark matter. These investigations are ongoing, and while they are bringing us closer to determining whether the AP has scientific merit, definitive proof remains out of reach – at least for now. With further advancements in observational technology, scientists may soon be able to gather the necessary evidence.

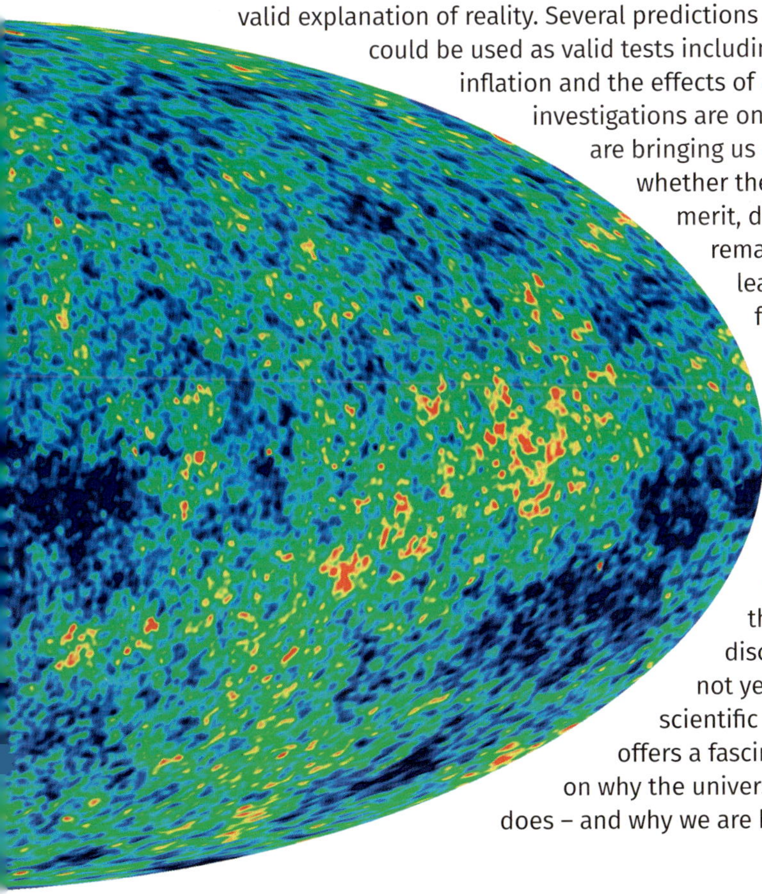

For the time being, the AP remains an intriguing hypothesis that continues to spark discussion. While it may not yet be provable in a strict scientific sense, it undeniably offers a fascinating perspective on why the universe appears the way it does – and why we are here to observe it.

FACT 33

There is a technique for spotting Dyson rings

Dyson spheres and rings have always held a special fascination for me. The concept is simple: build a great big structure, either as a sphere or a ring, to harness the energy from a star. The rings are far more simple and feasible to construct but, like their spherical cousins, Dyson rings remain, for now, a popular idea in science fiction; and yet they are starting to appear more and more in scientific debates.

The concept of the ring is similar to the sphere, a megastructure designed to encircle a star, harnessing its energy on a gargantuan scale. It might consist of a series of satellites or even habitats in a circular orbit with solar collectors. Unlike the spheres, the rings require far less resources to build. The concept of the sphere was first proposed by physicist and mathematician Freeman Dyson (1923–2020) in 1960. If such structures exist, they might be detectable and reveal the existence of intelligent civilisations.

It is fascinating to think about civilisations building such constructions. Our own civilisation currently consumes around 15,000 terawatts per hour and that number is only going to increase as our population grows and we become even more reliant on technology. To endeavour to quantify the energy usage of a civilisation based upon their level of technological capability, the Kardashev scale was developed. On this scale, we are currently at Type I which means our power usage is 4 × 1019 erg s-1 (4 terawatts.) If a civilisation requires 4 × 1033 erg s-1 (400 trillion terawatts) then it is considered to be type II and it is these civilisations that may be capable of building, and indeed may have to build, Dyson-like structures.

To create either a solid sphere or even a sphere with orbiting satellites would require phenomenal amounts of material. A sphere which had a radius of one astronomical unit (the average distance between Earth and the Sun) would require more material than exists in the entire solar system. It is far more likely that civilisations would create ring structures. Rings of this nature around a star would be able to harness significant amounts of energy but a ring around a pulsar would, if the pulse beam of the star could be tracked, be able to capture even more energy, of the order of 10 thousand trillion terawatts.

If – and again, it is a big if – Dyson-like structures exist, we now have a technique for spotting them! A team of scientists has worked out that by studying the light from a type of star known as a pulsar, Dyson rings could be found. Pulsars are stellar corpses emitting intense beams of energy that sweep across the cosmos like a great celestial lighthouse beam. If these beams strike a Dyson ring, then multiple images of the pulsar will appear simultaneously on the ring structure. Sounds easy – we just now need to find one!

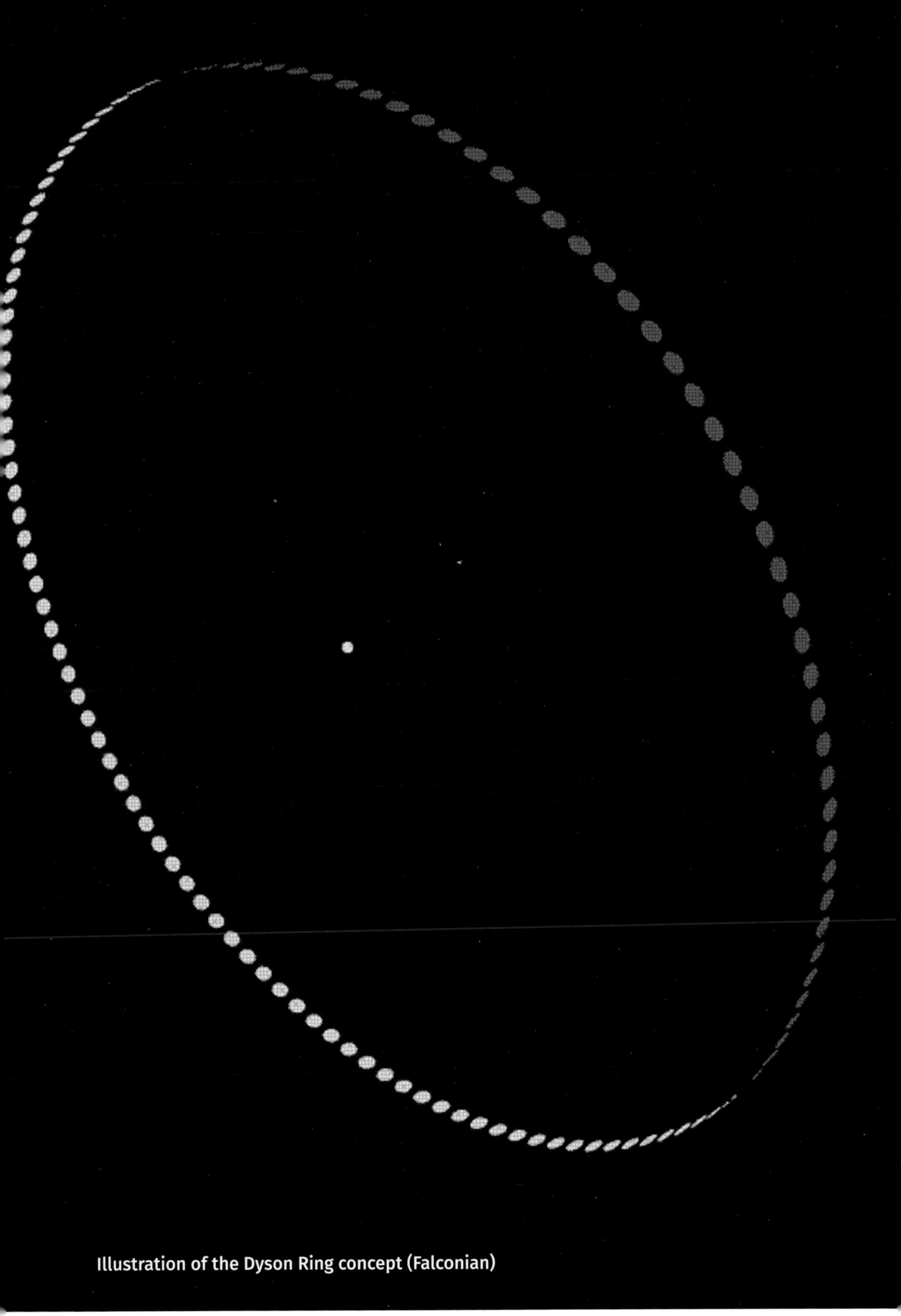

Illustration of the Dyson Ring concept (Falconian)

FACT 34

Pulsars are the ideal probes for dark matter

Pulsars are the remnants of massive stars that explode at the end of their lives in an event known as a supernova. As they rapidly spin, they emit a high-energy beam that sweeps across the cosmos, much like a lighthouse. When these beams align with Earth, they appear as regular pulses, making them some of the most precise cosmic timekeepers known. Discovered in 1967, pulsars have significantly advanced our understanding of stellar evolution and extreme astrophysical processes.

When a supermassive star reaches the end of its life, nuclear fusion in its core ceases. With the cessation of nuclear fusion, the outward force known as the thermonuclear pressure also ceases and so the outer layers collapse

The Lovell Radio Telescope is one of many telescopes used to capture data from pulsars. (University of Manchester)

inward, causing immense pressure that compresses the core to an extraordinary density. Through this process, the material that once formed the star is crushed into a sphere only a few tens of kilometres across, giving birth to a pulsar. Pulsars are closely related to neutron stars, as both form through the same process. The primary difference is that a pulsar emits a highly energetic beam, which periodically flashes across Earth, while a neutron star does not.

Recent observations of pulsars have revealed tiny variations in their precise, regular pulses, hinting at the presence of dark matter. These minuscule deviations in pulsar timing suggest that the radio waves emitted by the pulsar may be slightly redirected by an unseen mass between the pulsar and the telescope. The leading theory is that this undetectable mass is dark matter, an elusive substance that is thought to make up a significant portion of the mass of the universe.

By carefully analysing these timing delays and examining the precise arrival times of radio pulses – accurate to within a nanosecond – scientists can probe the regions between the pulsar and Earth. Data recorded from the Parkes Pulsar Timing Array (PPTA) survey, along with observations from telescopes such as Effelsberg, Nançay, Westerbork, Green Bank, Arecibo, Parkes and the Lovell Telescope in Cheshire, have been meticulously examined to uncover these subtle effects.

The results reveal that, when dark matter causes a delay, the arrival times exhibit distinctive patterns that correlate with the mass of the intervening object. Dark matter appears to slow the passage of light, affecting the pulsar signals received on Earth. A similar effect could be produced by normal matter, such as the Sun, but the key difference is that the object responsible for these delays is completely undetectable at any wavelength. If an object as massive as the Sun were causing the delays, the expected shift in timing would be around 10 microseconds. However, the timing discrepancies observed in the survey were 10,000 times smaller, suggesting the influence of a much subtler yet significant presence.

A detailed study of data from 65 'millisecond pulsars' has pinpointed approximately 12 pulsars where the effects of dark matter are evident. These findings bring scientists one step closer to unravelling the mysteries of dark matter and its role in shaping the universe.

FACT 35

There is a distinctive background glow of light across the universe

Think about background radiation. Most people immediately think of cosmic background radiation and story about pigeon excrement during its discovery. If you want to know about that story, stand by. It turns out though that the universe has several different types of background radiation, such as infrared and even gravitational wave backgrounds. But before we get into the different types, let us talk pigeon excrement.

Back in 1964, two scientists were exploring signal interference at the Bell Laboratory, New Jersey. Physicist Arno Allan Penzias and radio-astronomer Robert Woodrow Wilson detected a faint 'hum' in all directions and initially put it down to pigeon excrement as these birds nested in the horn of the radio receiver. Instead, what they had discovered was cosmic background radiation, the faint glow that permeates the entire universe and is the thermal radiation

The *New Horizons* spacecraft used to capture the cosmic optical background. (NASA)

left over from the Big Bang. Studying it allows us to understand more about the universe when it was 380,000 years old.

In the late 1980s a different type of background radiation was detected: infrared background radiation. It consists of the diffuse infrared glow that fills the universe and comes from numerous sources across the cosmos, mostly thermal emissions from dust grains heated by stellar radiation. In addition to this is gravity wave background, although this has yet to be detected.

Another hotly debated background is the cosmic optical background (COB), a diffuse light which originates from stars and galaxies and spans the whole of the visible spectrum. There has been gathering momentum in its study however, with observations from the Hubble Space Telescope and the Spitzer Infrared Telescope (now known as the Spitzer Space Telescope). These studies have revealed that a large contribution to a general background optical glow come from faint unresolved galaxies. Studying the COB allows us to explore the total energy output of the universe, helping us to understand galaxy and star formation across space and time.

The detection of the COB is a challenging one, however, with Earth-based instruments or even those in Earth orbit plagued by interference. 'Zodiacal light', for example, is the result of sunlight scattered by interplanetary dust, and is dominant in the inner solar system, making studies of the COB difficult. The *New Horizons* probe is ideally positioned beyond the orbit of Pluto, over 8 billion kilometres away from any human or natural interference. On board *New Horizons* is the LORRI (Long Range Reconnaissance Imager) camera which was identified as an ideal platform to begin a search.

Using images from the LORRI camera, a team of astronomers attempted to measure the COB over the range 0.4 to 0.9 micrometres. The images were from high galactic latitudes to ensure no diffuse light from the Milky Way or scattered light from bright stars. Isolating the COB contribution to the total sky brightness levels required digitally subtracting the scattered light from bright stars and galaxies and from faint stars within the field that were fainter than that detectable by LORRI. Interestingly, the results showed that, based on the estimated galaxy counts in the sampled regions of the sky, there was indeed an optical glow and that glow is the result of light from all the galaxies within our observable region of the universe.

FACT 36

Oxygen in deep alien oceans of water worlds can sustain life

I have always found the Mariana Trench fascinating. It is like an alien world right on our doorstep. Any visitor to the oceans or seas of our planet will hopefully get to see fish flitting around and while they can survive in this alien underwater world they still need oxygen. Breathing in oxygen is a familiar experience to us – we inflate our lungs and suck air into them to keep us topped up with life-giving oxygen. Fish are different – they get their oxygen as water flows over their gills.

The production of oxygen by plants is well understood. Light is captured by a pigment known as chlorophyll where it is then converted into chemical energy and stored in the glucose. During photosynthesis, carbon dioxide from air and water from soil are combined in a series of chemical reactions to produce glucose and the oxygen we breathe. This oxygen from the plants plays a role in maintaining levels in the atmosphere and the oceans and seas. However, recent studies have challenged this somewhat simplified explanation.

Thanks to robotic underwater explorers the sight of life teeming around thermal vents on the bottom of the deep ocean is not unusual. At those depths, no sunlight can penetrate to facilitate photosynthesis in plants. Somehow though, oxygen is present in the dark, deep, gloomy regions of the ocean floor and it is the rocks that are the secret to the unusual presence of oxygen. Surprisingly, water samples from the deep showed a rise in oxygen levels instead of an anticipated decline.

Lab tests ruled out the possibility of microbes creating the oxygen but the region being explored was peppered with lumps of rock known as polymetallic nodules. The nodules are known to form when manganese and cobalt precipitate out of water and form around shells. The nodules where theorised to be the source of the oxygen but the mechanism was not understood.

It is known that putting batteries in salt water results in bubbles of hydrogen and oxygen being produced, the result of a process known as *electrolysis*. A team studying the problem looked at the rocks at the bottom of the ocean and measured the voltage on the nodules, with just one of them producing 0.95 volts, a little lower than the required 1.5 volts for saltwater-driven electrolysis, but the team were onto something, suspecting multiple rocks could cluster together to increase the voltage and generate the oxygen.

The discovery of rocks on the bottom of the ocean generating oxygen is fascinating on its own but it has profound impacts on the search for life elsewhere in the universe. We have already discovered ice-covered water worlds among the moons around some of the outer planets. It is likely there will be others in planetary systems around other stars. If these worlds are common then it is quite likely that oxygen is being released through electrolysis from similar metallic nodules and perhaps supporting entire ecosystems.

Hydrothermal vents at the bottom of the ocean. (NOAA)

FACT 37

A star (possibly) has been discovered and it is escaping the Milky Way

Citizen science is such a great concept. Using the combined computing power of a gazillion (exaggeration) desktop and laptop computers to churn through data is an excellent and efficient way of analysing volumes of information. It is not a new idea though, since the concept of using members of the public to support scientific endeavours dates back further than you might think. One of the earliest examples can be seen at the beginning of the twentieth century when the 'Christmas Bird Count' was launched by the Audubon Society in 1900. It encouraged members of the public to track bird populations. When it comes to space science projects, 'SETI@Home' springs to mind, among others.

Another citizen science project, 'Planet 9', was launched by Backyard Worlds in 2017 and to date it has 82,690 volunteers signed up. The goal is to hunt for possible distant planets in our solar system, or previously unidentified brown dwarfs in our solar neighbourhood. It takes images from NASA's Wide-field Infrared Survey Explorer (WISE) and shares them with its volunteers who then study them for objects that are moving.

While hunting for objects worthy of study, a team of volunteers identified a fast-moving object which became known by the catchy title CWISE J124909.08+362116.0. Observations were scheduled after its discovery by several ground-based instruments that showed it was moving so fast that it will eventually escape the gravitational pull of the Milky Way, heading out to intergalactic space. It is the first object with a mass smaller than or similar to that of a star that has been seen travelling at such eye-watering speeds, around 1.5 million kilometres per hour!

Quite what CWISE J1249 is remains a little bit of a mystery at the moment. It has a low mass suggesting either a low-mass star, or possibly a gas giant planet, or maybe something in between known as a brown dwarf. And yet it has also been found to have less iron and other metals in the core than other stars and brown dwarfs. The composition suggests it may be quite old and perhaps one of the first generations of stars from the Milky Way. If it is a brown dwarf then in itself it is not particularly rare. 'Planet 9' has discovered more than 4,000 of them alone, but none are travelling at such a great speed that they will escape the Galaxy.

It is not just the nature of the object that is perplexing astronomers, it is its high speed that has been the subject of much debate. One theory suggests it could have been a part of a binary system with a white dwarf. The white dwarf would slowly accrete material from its companion leading to an explosion when the material reached a critical point. This event may have provided the energy to accelerate the object out of the system and beyond. Alternatively it may have been a part of a cluster which was ejected through gravitational interactions.

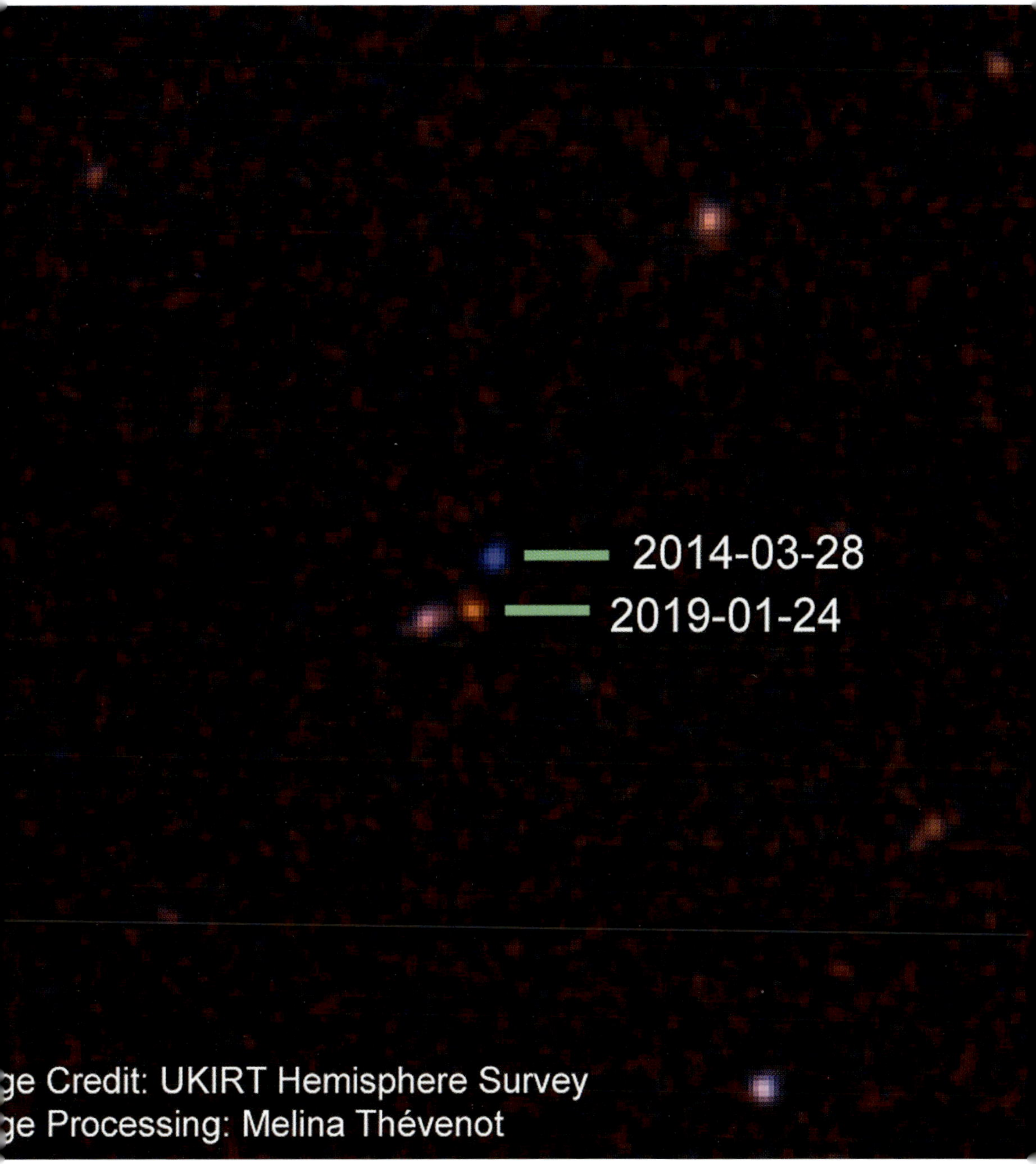

CWISE J1249, a suspected star escapting the Milky Way Galaxy (UKIRT)

FACT 38

Gravitational wave observatories can detect collapsing stars

The merging of black holes and neutron stars is among the most energetic events in the universe. Not only do such events emit colossal amounts of energy, they can also be detected through gravitational waves. Observatories like LIGO (the Laser Interferometer Gravitational-Wave Observatory), Virgo and KAGRA have detected their gravitational waves but new gravitational wave observatories are now thought to be able to detect the collapse of a massive rapidly-spinning star before it becomes a black hole.

The acceleration of massive objects can create ripples in space-time known as *gravitational waves*. They were first predicted by Albert Einstein in 1915 in his general theory of relativity. The waves are thought to travel at the speed of light and carry energy across the cosmos. Unlike electromagnetic waves, the gravitational waves seem to interact weakly with matter, which allows them to pass unimpeded through stars, planets and galaxies. In 2015, the first gravitational waves were detected by LIGO.

Black holes and neutron stars are the remains of the death of stars. When supermassive stars reach the end of their lives they create astronomically (pardon the pun) dense objects. Neutron stars are stellar cores where the space between neutrons has been squeezed out, leaving behind one great big neutron, often just a few tens of kilometres across. The remains of even more massive stars get compressed to an object of infinitely small size, a *singularity*, the power house of a black hole.

As the lives of these massive stars ends, the core runs out of fuel and no longer generates the thermonuclear force to stop the collapse. The collapse leaves behind a large disc that quickly spirals around before falling into the black hole and it is this, it is thought, that generates gravitational waves.

To date, merger events have only been observed through gravitational waves, however a team of researchers has run simulations and showed that collapse events produce gravitational waves powerful enough to be detected from a

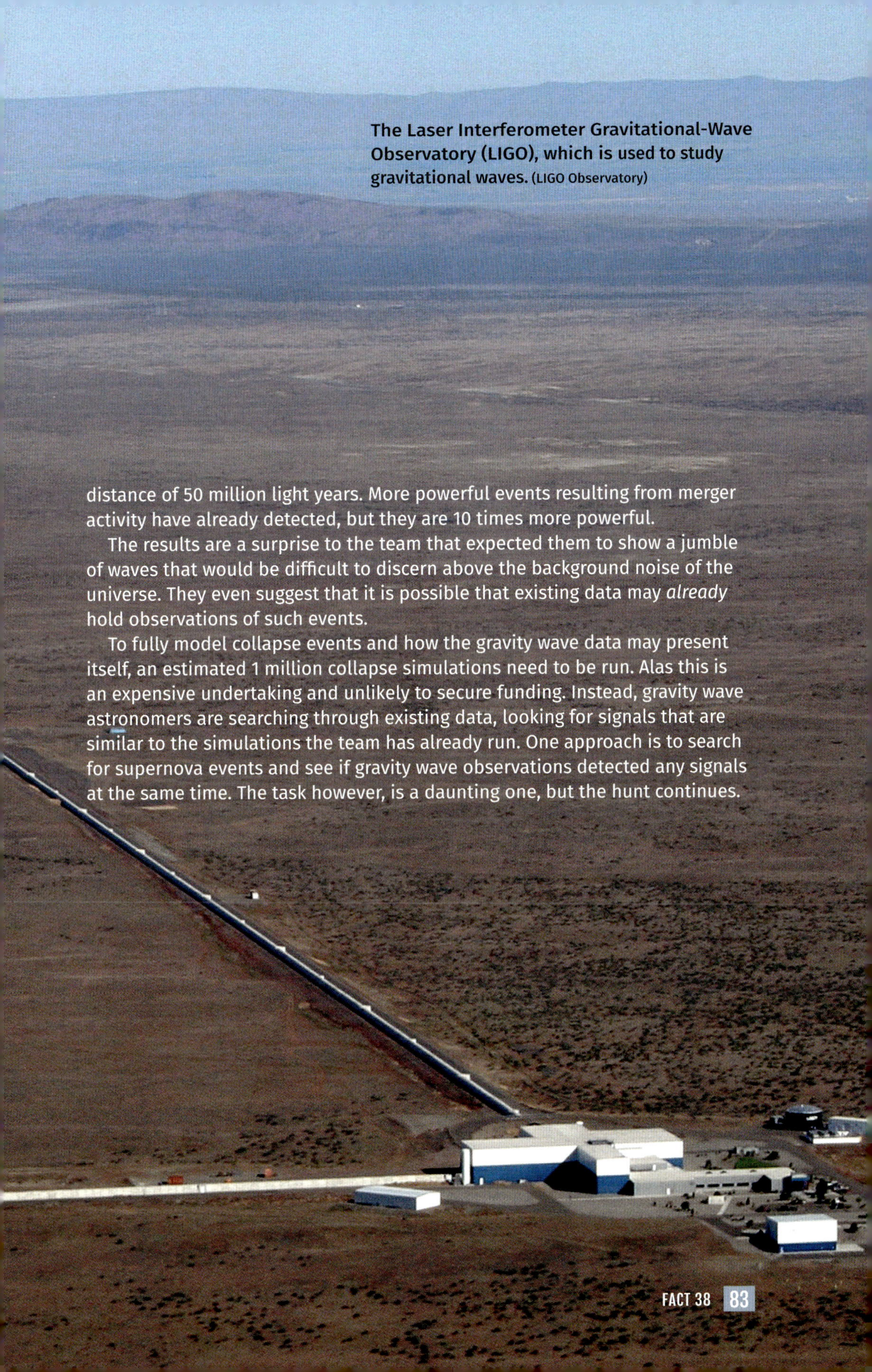

The Laser Interferometer Gravitational-Wave Observatory (LIGO), which is used to study gravitational waves. (LIGO Observatory)

distance of 50 million light years. More powerful events resulting from merger activity have already detected, but they are 10 times more powerful.

The results are a surprise to the team that expected them to show a jumble of waves that would be difficult to discern above the background noise of the universe. They even suggest that it is possible that existing data may *already* hold observations of such events.

To fully model collapse events and how the gravity wave data may present itself, an estimated 1 million collapse simulations need to be run. Alas this is an expensive undertaking and unlikely to secure funding. Instead, gravity wave astronomers are searching through existing data, looking for signals that are similar to the simulations the team has already run. One approach is to search for supernova events and see if gravity wave observations detected any signals at the same time. The task however, is a daunting one, but the hunt continues.

FACT 39

Iron winds are blowing on WASP-76b

Exoplanets exist outside of our solar system and orbit other stars. The first confirmed discovery was back in the 1990s and since then over 5,200 have been discovered. Many of them are gas giants like Jupiter or Saturn and others are small, rocky Earth-like planets, minus perhaps their habitability status. Some lie in habitable zones where liquid water could potentially exist, fuelling hopes of finding life. As more advanced telescopes and detection techniques are developed, not only will our detection levels increase further, but so will our ability to explore these alien worlds in greater depth and detail.

One such exoplanet, WASP-76b, has received quite a lot of attention of late. It is an ultra-hot gas giant that is 640 light years from us in the direction of the constellation Pisces. It was discovered back in 2013 and has an orbit that is very close to its host star, completing one orbit in just 1.8 Earth days! It is the proximity to the star that has led to the extreme daytime temperatures of over 2,000 degrees. The intense heat is thought to vaporise iron, which then

Exoplanet WASP-76b. (NASA)

condenses into liquid on the cooler night-time side and falls as iron rain – a phenomenon both fascinating and terrifying.

Generally speaking, exoplanets have been discovered with a wide range of environmental conditions. WASP-76b, however, is one of the most extreme, with its intense daytime temperature. It is a world that is tidally locked to its host star, so intense winds encircle the planet. These winds, which can reach thousands of kilometres per hour, contain high quantities of iron atoms that stream from the lower to upper layers around the atmosphere, creating a complex dynamic system unlike anything seen in our own solar system.

Astronomers have used the ESPRESSO spectrograph that was installed on the European Southern Observatory's (ESO) Very Large Telescope (yes, that is its name!). It is known for its stability and high spectral resolution, so it can discern wonderfully fine levels of detail in a stellar spectrum, making it ideal for exoplanet research.

Using a technique known as high resolution emission spectroscopy, the star and planetary system was studied in the visible light spectrum. This approach relies upon the detection of emission lines in a spectrum and enables the chemical composition to be decoded. Here, they detected the chemical signature of iron and found that WASP-76b's iron winds were moving from lower levels to the higher layers of the atmosphere, possibly driven by strong thermal gradients.

The study of exoplanet atmospheres helps us to further develop our understanding of the range of environments on these alien worlds. As a gas giant, the discoveries on WASP-76b help us learn a little more about the climates of worlds that are barraged by extreme levels of radiation from their host star. The continued exploration of such exoplanets not only deepens our knowledge of planetary science but also sheds light on the diversity and complexity of planets beyond our own system.

FACT 40

Slime can teach us about the cosmic web

The cosmic web is the largest scale structure of the universe and it is made up of filaments of galaxies and dark matter that stretch across the gulf of space. The filaments connect galaxy clusters with immense voids in between. The web-like structure has formed as a result of the force of gravity pulling matter together since the beginning of time. Studying the cosmic web helps us to piece together the evolution of the universe, how matter is distributed and its relationship with dark matter. It acts as a kind of scaffolding on which galaxies form, grow and interact, revealing clues about the fundamental forces that shape the cosmos.

Since the early 1980s, it is been known that the nature of a galaxy and its environmental properties has an impact on how it grows and evolves. The exact nature of how this happens is still the cause of many debates. A team of researchers believe they may have demonstrated how galaxies evolve using a surprising new tool. Computers truly are wonderful things and powerful – but only if they are programmed by a skilful mind. There is an algorithm that mimics the growth of slime mould, but a team of researchers have adapted it to model the large-scale structure of the universe, bridging the gap between biological systems and astrophysics.

There are similarities, albeit abstract in some ways, between the structure of slime mould and the cosmic web. The algorithm used to model slime mould had to be modified to help visualise the cosmic web, but advanced graphics modules meant it could be repurposed to visualise the universe. By comparing the paths taken by slime mould to find food with how matter flows through the universe, scientists created a new way to trace cosmic structures. The algorithm's original purpose was designed to mimic slime mould that could find its own food by reforming itself into a structure much like the cosmic web. It took several years, but the algorithm has now been repurposed!

The result produced far more detailed discrete structures than previous methods. It even went on to produce hypotheses of changes to galactic structure and galactic evolution – suggesting, for example, that in earlier epochs, the growth of a galaxy was stimulated by proximity to larger structures. In the near universe, and therefore in cosmologically recent times, we see that galaxy growth is limited by that same proximity. This was not possible without the modified slime mould algorithm. We can now map out the gas around the real universe using the algorithm across many different times to help understand how the web has changed and the universe evolved. This opens the door to more accurate simulations and a deeper understanding of cosmic history, helping astronomers make sense of the vast and complex universe we inhabit.

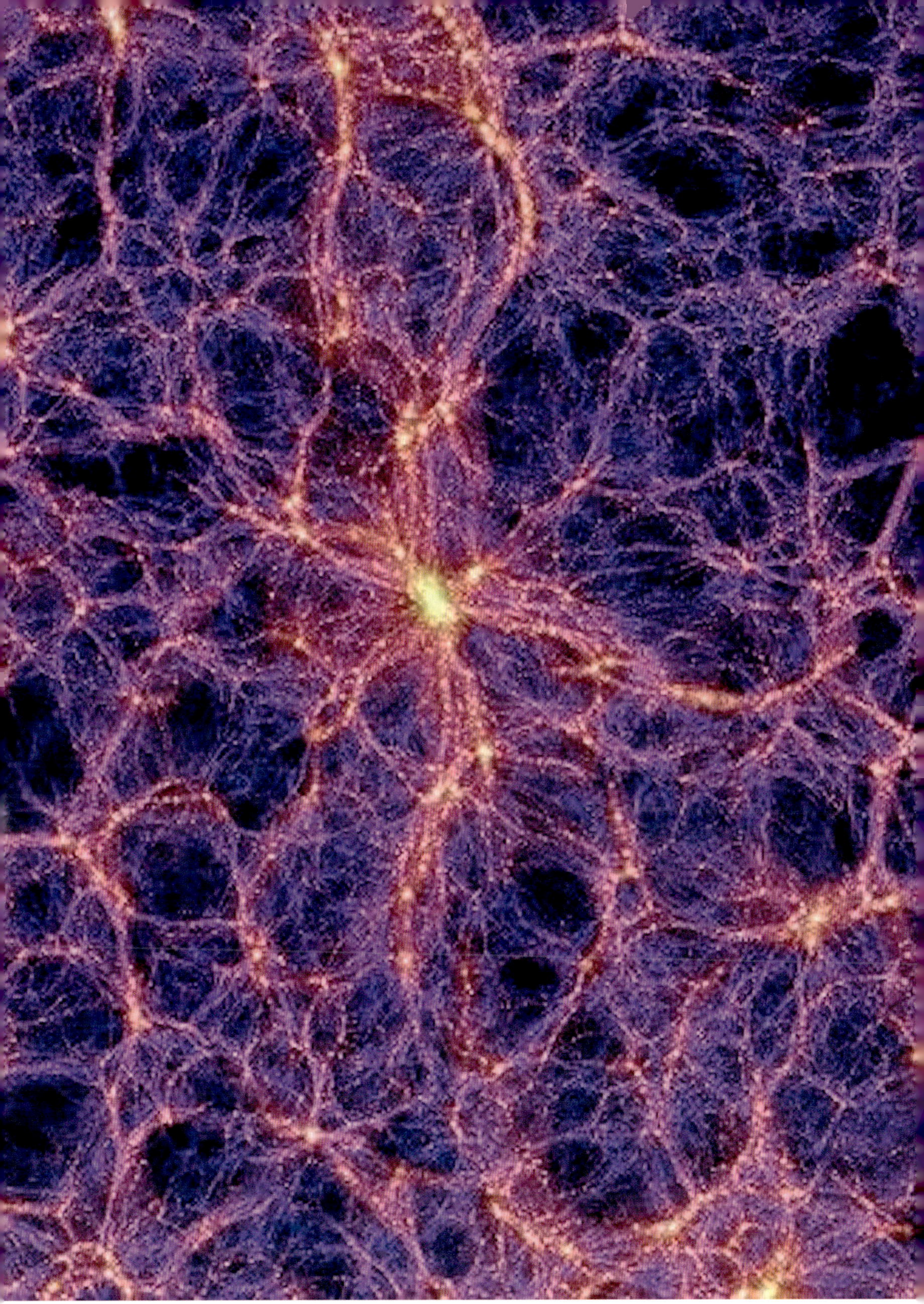

The cosmic web.

FACT 41

The longest black hole jet is 23 million light years

Mass in the universe creates distortions in space-time, with greater mass producing stronger distortions. Black holes represent extreme gravitational regions where not even light can escape. They form through the collapse of massive stars that have depleted their core fuel, creating a point of infinite density called a *singularity*. The *event horizon* surrounds this singularity at a distance determined by the original star's properties, and anything crossing this boundary is permanently trapped.

Black holes generate powerful jets: high velocity streams of particles ejected from their vicinity. This material never actually reaches the event horizon but instead is expelled from within the accretion disc. The rotation of these discs of heated matter, combined with the black hole's magnetic fields, can produce jets from the polar regions. These jets can approach light speed and extend across vast distances of space, though their exact formation mechanisms remain incompletely understood.

Observations using the Low-Frequency Array (LOFAR) radio system have revealed a remarkably large jet spanning the equivalent of 140 Milky Way galaxies aligned side by side. For context, the jet from Centaurus A at our Galaxy's centre spans approximately 10 Milky Ways. This massive structure, nicknamed Porphyrion after a Greek mythological giant, dates to when the universe was 6.3 billion years old and generates power equivalent to trillions of suns.

Astronomers theorise that such enormous jets, if common in the early universe, may have significantly influenced galaxy formation. Contemporary jets observed in the nearby universe appear considerably smaller, suggesting that these ancient giant jets may have connected to neighbouring galaxies, supplying energy and material that drove their development.

The LOFAR survey identified more than 10,000 of these megajets, substantially more than previous studies that found only hundreds, indicating they were more common than previously thought. The research required extensive effort, with the team examining radio images visually, employing machine learning tools, and enlisting citizen scientists worldwide to assist in the identification process.

This discovery has significant implications for our understanding of cosmic evolution. The energy transfer from these jets may have helped regulate star formation in early galaxies and distributed heavy elements across vast regions of space. The identification of so many megajets challenges previous models of black hole behaviour and suggests that these phenomena played a more central role in universal development than previously recognised.

The jet emanating from Centaurus A. (ESO/WFI)

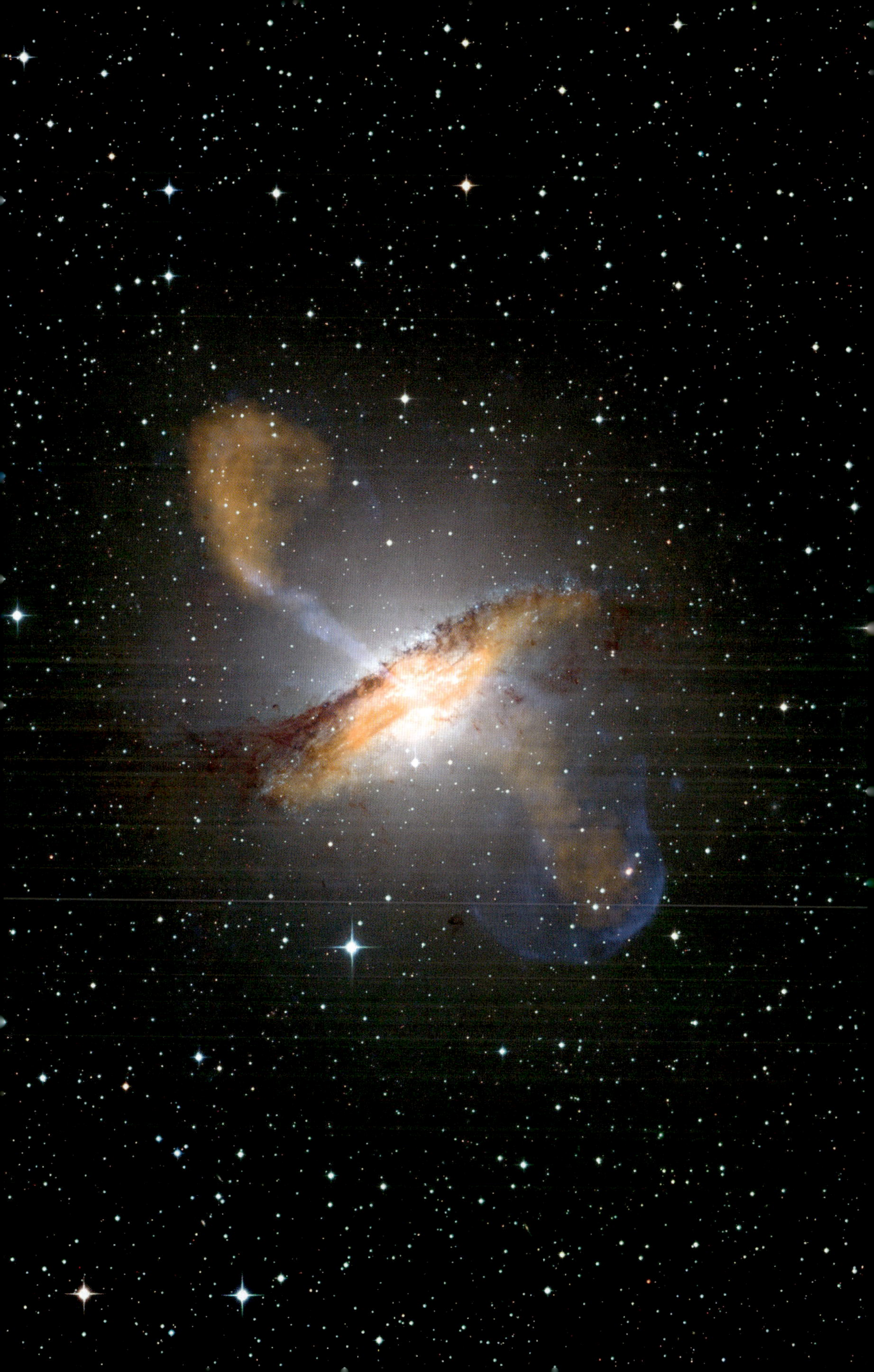

FACT 42

Scientists have simulated nuclear detonations in space

Statistically, the risks of an asteroid hitting Earth are low, but the 'impact' of such an event could be catastrophic. Most asteroids that enter our atmosphere burn up, giving us the stunning sight of a 'shooting star', but those over 1 kilometre wide could cause widespread damage and devastation. The likelihood is rare and might occur once every several hundred thousand years, but smaller objects hit more often. They can also create significant localised

The DART spacecraft under construction. (NASA)

damage. Take the Chelyabinsk Event in Russia in 2013 when an asteroid exploded in mid-air, sending shockwaves across hundreds of kilometres.

While the likelihood is low, we need a plan to deal with such threats when they arise. The Double Asteroid Redirection Test (DART) mission that NASA launched back in 2021 sent a probe to the binary asteroid system Didymos with its tiny moon Dimorphos (See Fact 12). The probe hit Dimorphos in September 2022 and slightly altered the orbit, proving it is possible to change an asteroid's trajectory. Though this approach worked, it has limitations since ramming a spacecraft into them may not be effective on large asteroids. Combined with the likelihood of not getting much notice, we need a more effective alternative.

Scientists have looked at various options – from attaching fusion engines to the space rocks (asteroids), focusing laser beams on them, neutron bursts and nuclear blasts that generate X-ray radiation. After analysing these options, only nuclear blasts have been deemed suitable for neutralising the threat of a large asteroid impact when time is limited.

Simulations have been run using a nuclear bomb to deflect an incoming asteroid. A good chunk of the energy from a nuclear explosion comes out as X-rays. The simulations showed that these X-rays would be powerful enough to vaporise the asteroid's surface, with the resulting vapour slowly pushing the asteroid in the opposite direction. Think of it as a basic rocket engine with the vapour creating thrust. In simulations, the test asteroid reached speeds of 250 kilometres per hour!

The results showed that X-rays could work and might provide sufficient protection against an incoming asteroid up to 4 kilometres wide – assuming, of course, we spot it early enough! There's the challenge – asteroids are typically dark and spotting them against the blackness of space is not easy. The more time we have, the better our chances of deflection working.

The next step is actual testing, but nuclear explosions come with serious costs, risks and a ton of international legal restrictions. We need careful planning and probably more research before this approach can be ready for use if needed. Recent funding increases for planetary defence programmes and improved detection systems have made this work more urgent and feasible than ever before.

FACT 43

A black hole has been seen eating a star

Black holes are stellar corpses with such intense gravity that nothing, not even light, can get away. They form when big stars collapse at the end of their lives, crushing down to an infinitely tiny point called a *singularity*. Forming around the singularity is the *event horizon*, the ultimate point of no return. Even though they are super hard to see directly, we can spot black holes by watching how they mess with nearby stars and gas clouds. There's still a lot we do not know about black holes, so scientists study them intently to learn more.

A group of astronomers analysed data from the Catalina Real-Time Transient Survey (CRTS) to check out short-lived space events. The CRTS started in 2004 and scans huge patches of sky looking for things that change, like supernovae and asteroids. It uses telescopes in Arizona to hunt for brief cosmic happenings and has been fabulous for helping learn about how stars live and die and what distant galaxies are up to.

The team spotted a super-bright flash in a galaxy about 500 million light years away and it all happened in a tiny galaxy, about 400 times smaller than our Milky Way. The burst, named CSS161010, got super-bright in just four days, then dimmed to half brightness 2.5 days later. Later they realised the All-Sky Automated Survey for Supernovae (ASAS-SN) had actually caught it earlier too. Luckily, they spotted it early enough that other ground telescopes could swing into action and take a look. These events are usually tough to study because they are usually transient in nature and short-lived.

Only a few events like CSS161010 have been caught in recent years, and until now nobody really knew what they were. The team analysed the light patterns and found hydrogen moving at speeds up to 10% as fast as light itself. The way the hydrogen emissions changed looked a lot like what happens around active galactic nuclei where huge black holes are found. This suggested the flash is related to a black hole, but not one of the supermassive ones.

The object got 900 times dimmer over the next two months. When they looked at the spectrum again, they still saw those blue shifted hydrogen lines showing gas shooting out super-fast. This is not something you typically see with supernovae, which points to something else going on. The conclusion is that it is a black hole devouring a star.

This discovery represents a significant advancement in our understanding of *tidal disruption events*, where stars are literally torn apart by black holes. The rapid energy release and distinctive spectral signature could potentially provide astronomers with a new method for identifying and cataloguing smaller black holes throughout the cosmos, filling a crucial observational gap in our knowledge of these fascinating cosmic predators.

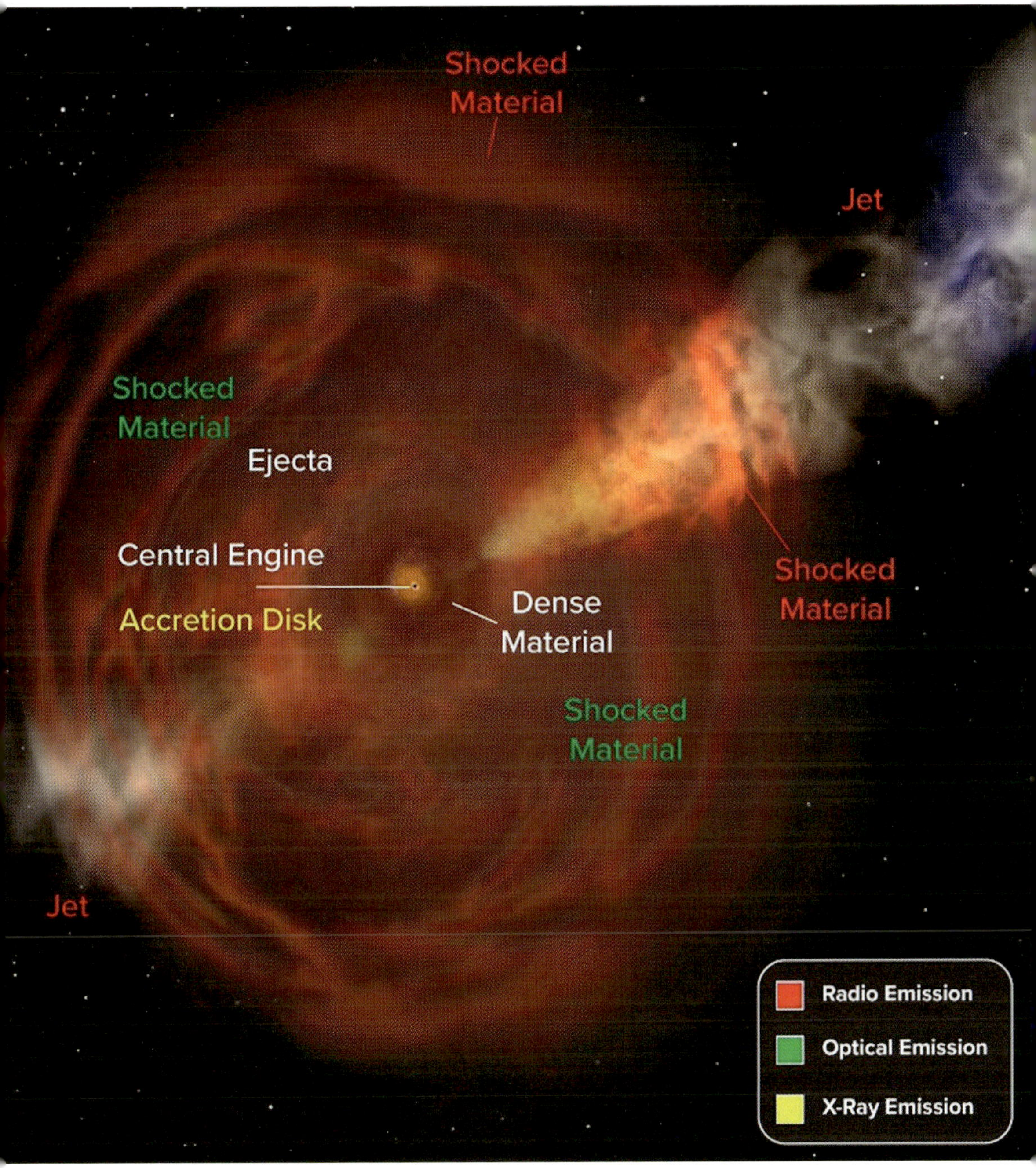

Illustration of the processes behind CSS161010 (Bill Saxton)

FACT 44

Stars can get hiccups

A supernova marks the end of the life of a massive star. The event is one of the most energetic processes in the universe, releasing immense amounts of energy. There are two types of supernova: Type I occurs in a binary star system and Type II at the end of a star's life. Stars that are more than eight times the mass of the Sun run out of nuclear fuel and, suddenly, the outward pressure generated by fusion ceases. The core collapses under gravity causing a violent explosion that ejects the outer layers of the star.

The process is an essential step in the evolution of life, since all the heavy elements needed to form life have been synthesised inside massive stars and it is the supernova process that liberates them to spread throughout the universe. What remains is dependent on the mass of the progenitor star and will either be a neutron star or black hole.

Before the star 'goes supernova' however, there has for some time been a theorised phase during which the star undergoes what has been described as the 'hiccups'! Until now though, they have just remained a theory. This phase is perhaps even more rare than a supernova, happening very infrequently and only to exceptionally large stars between 60–150 times the mass of the Sun.

Observations finally proved stars *can* hiccup, a process called 'pulsational pair-instability' (PPI). The rarity of these events and the relative faintness of the hiccup is what has made them hard to detect, until now! In December 2020, a team detected a supernova (SN2020acct) in a galaxy called NGC2981 and, as expected, the light from it faded. Two months later, they detected light from the same region of the galaxy – that's unusual since it is very rare for a Type II supernova to repeat itself.

Further study revealed the supernova the team had thought they had detected was light being produced by slow-moving shells of material colliding near the star. It was not a supernova. It turns out the second burst of radiation was the supernova, the first was one of the first observations of a star suffering from hiccups!

In massive stars, the core develops to a very high temperature which contracts and expands in rapid succession. This might occur in the last few years, or even days, the timescales are still not clear. Each time the stellar core pulsates, a shell of material is ejected causing the star to slowly lose mass. On occasions, the ejected shell of material collides with other shells, creating the intense burst of energy that we have now been able to see as 'hiccups'.

Supernova 1987A pictured here at the centre, near the Tarantula Nebula. (ESO)

FACT 45

You are made of carbon that took a journey into intergalactic space

Carbon is one of the key building blocks of our bodies. But if you go back to the very beginning – around 13.7 billion years ago – the universe was mostly made up of hydrogen, with just a hint of helium. The carbon we carry around today was not there at the start; it was forged inside stars through a process

The Cosmic Origins Spectrograph (COS) before being fitted to the Hubble Space Telescope. (NASA)

called *nucleosynthesis*, the cosmic alchemy that creates the heavier elements essential for life as we know it.

As stars age, they go through a series of fusion reactions. They start by fusing hydrogen into helium, and later on helium nuclei combine to form carbon. It takes three helium-4 atoms to make one carbon-12 atom, and for that to happen, temperatures need to hit around 100 million kelvin. As these reactions take place, they generate an outward push known as *thermonuclear force*. During the main phase of a star's life, this push is perfectly balanced by the inward pull of gravity, keeping the star stable for millions or even billions of years.

Eventually, depending on the star's mass, it reaches a point where it cannot hold onto its material any more. That material, rich with heavier elements like carbon, gets blown out into space in spectacular events like supernovae or stellar winds. Some of it ends up in 'stellar nurseries' – those regions where new stars, planets and maybe even life can begin to form.

But those carbon atoms do not just float around aimlessly until they find their next home. In galaxies like the Milky Way, where stars are still being born, the journey is much more dynamic. Huge flows of gas called the *circumgalactic medium* swirl around the galaxy and stretch way out into intergalactic space. These currents pull the stellar material outward and then gradually draw it back in, helping seed new generations of stars.

Thanks to the Cosmic Origins Spectrograph (COS) aboard the Hubble Space Telescope, astronomers have been able to study how these carbon atoms interact with the circumgalactic medium in 11 star-forming galaxies. What they found was fascinating: the carbon absorbs light in specific ways, revealing its presence far from its original star. In one case, carbon was detected nearly 400,000 light years from its galaxy – about four times the width of the Milky Way!

It is a pretty astonishing discovery. Instead of gently drifting through space, these elements go on a wild, cosmic ride. And while this research has already reshaped our understanding of how stars and galaxies evolve, there's still much more to learn. As we dig deeper, we'll get a clearer picture of how galaxies grow, why some keep making stars, and why others become barren stellar wastelands.

FACT 46

Scientific models now support the potential stability of Dyson spheres

The Dyson sphere concept (see Fact 33), proposed by physicist Freeman Dyson (1923–2020) in 1960, proposes a theoretical megastructure designed to capture a star's energy output. Such a structure might manifest as either a complete shell or a network of solar collecting satellites orbiting a star (a Dyson ring), potentially providing vast amounts of energy. Dyson recognised that constructing a solid sphere would need significant volumes of material and face stability challenges in construction, suggesting a satellite swarm as a more practical alternative.

Recent research has generally dismissed the solid sphere design in favour of satellite configurations. This scepticism aligns with the findings of physicist James Clark Maxwell (1831–1879) in 1856 regarding Saturn's rings, which demonstrated that such structures could not maintain stability as solid, uniform entities due to gravitational interactions.

However, the model of the 'circular restricted three-body problem', a fundamental concept in celestial mechanics, can be used to examine the motion of a negligible mass object under the gravitational influence of two larger bodies in circular orbits around their common centre of mass. Within such systems exist five equilibrium points, known as Lagrange points, of which two (L4 and L5) can provide stability under specific mass-ratio conditions, such as those between Jupiter and the Sun.

It seems that certain configurations could indeed support stable spherical or ring structures. One viable arrangement occurs when two primary masses orbit their common centre of mass while a uniform ring encircles the smaller mass. More notably, even a complete sphere could achieve stability if positioned to enclose the smaller of the two masses.

These results suggest intriguing possibilities for Dyson sphere implementation beyond the realms of science fiction. The findings indicate that stellar systems throughout the cosmos might potentially host such structures, constructed by advanced civilisations to harness stellar energy.

The implications are beyond theoretical astrophysics and take us into the realm of astrobiology and the search for extraterrestrial intelligence (SETI). If stable Dyson structures are indeed possible, they might represent one of the most distinctive technosignatures potentially detectable across vast cosmic distances.

Astronomers have already conducted preliminary searches for the infrared signatures that Dyson structures might produce, as they would absorb visible light from their host star and re-emit it as infrared radiation. The identification of suitable binary star systems could provide astronomers with promising targets for future observation campaigns. While the engineering challenges remain formidable, this research reopens possibilities for megastructure designs previously considered impossible, challenging us to reconsider the upper limits of technological advancement in the universe.

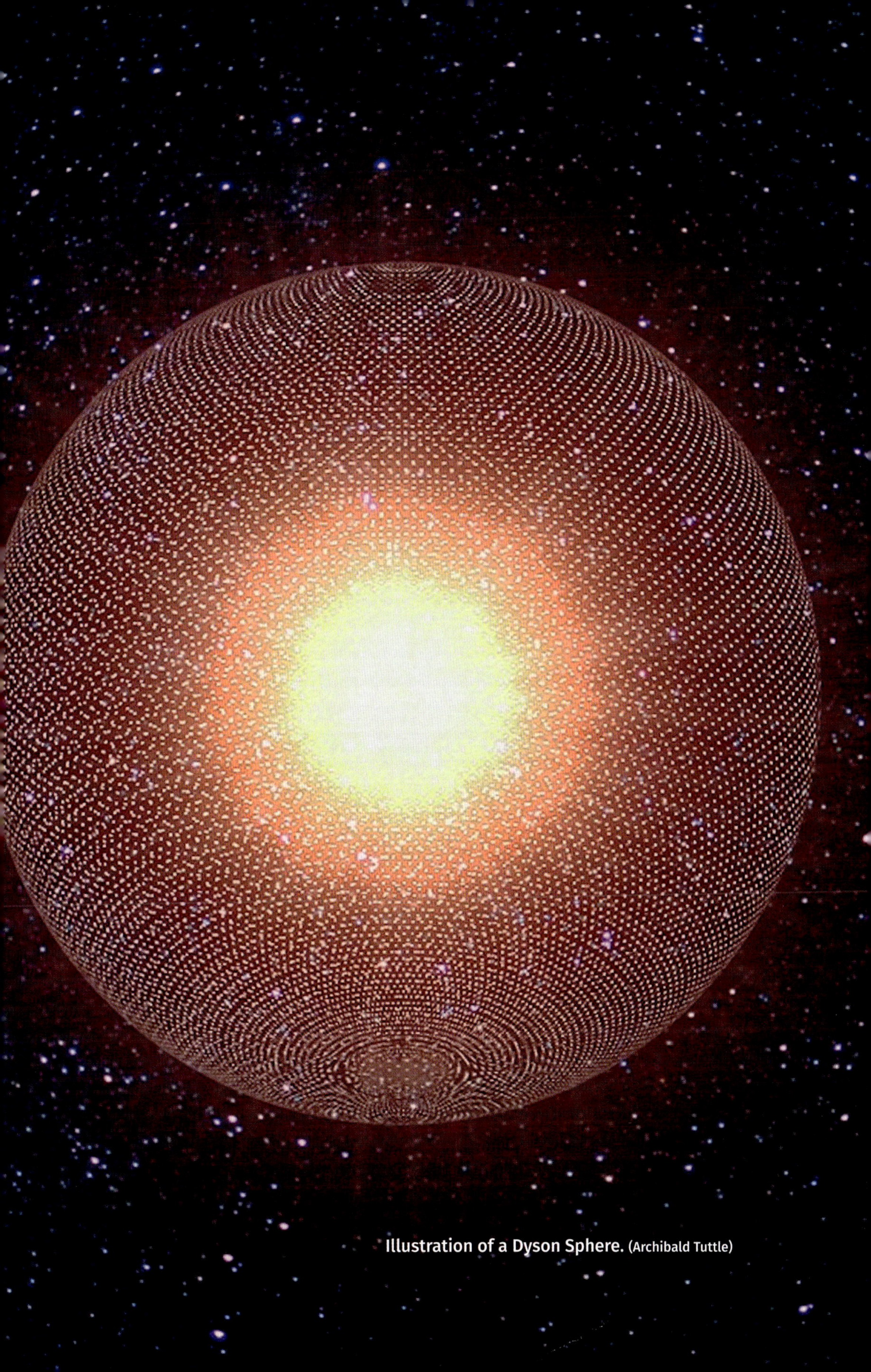

Illustration of a Dyson Sphere. (Archibald Tuttle)

FACT 47

Runaway stars can escape the Milky Way

High velocity stars travel through space at extraordinarily high speeds, often between 500 to 2,000 kilometres per second, compared to the Sun, which travels at a mere 200 kilometres per second. These rapidly moving stars are usually expelled from their galaxies due to gravitational forces, often from

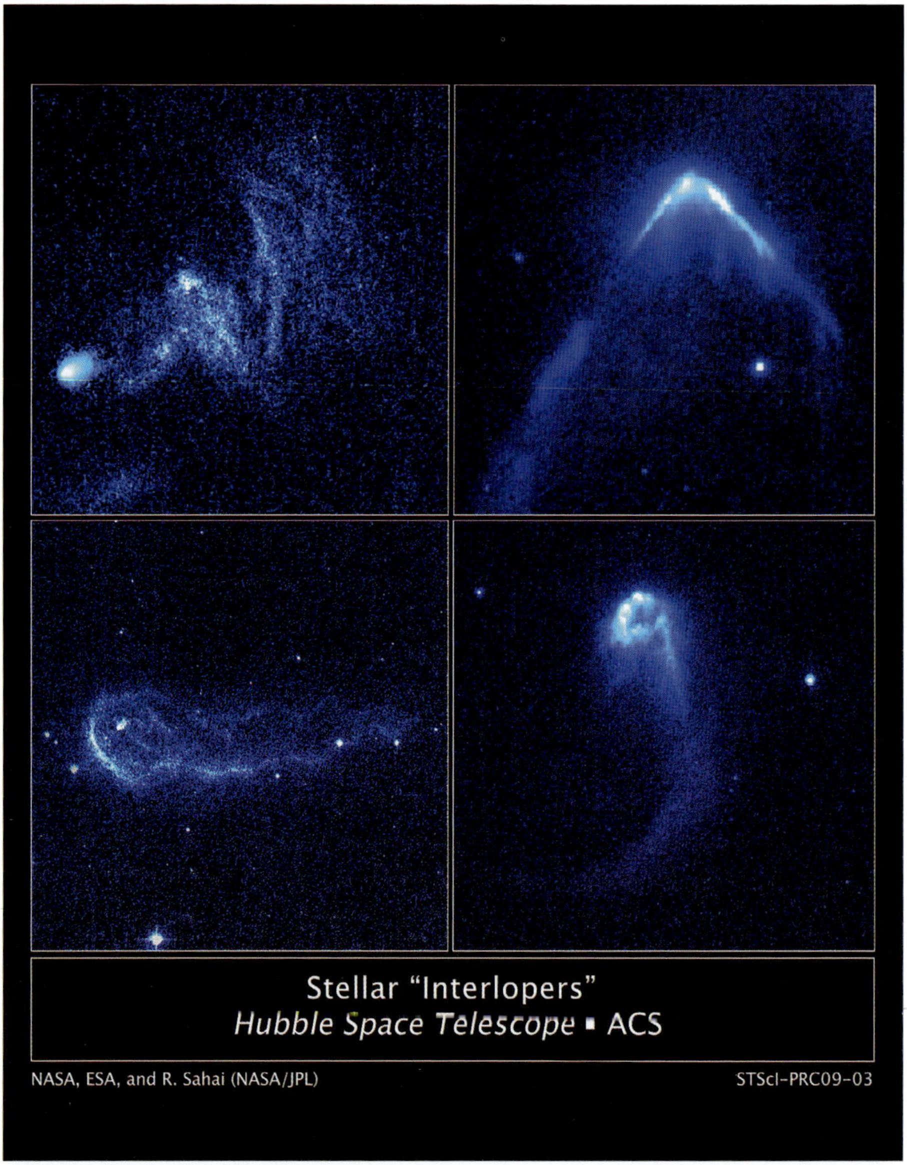

close encounters with supermassive black holes or other stars. Some move so fast that they can break free from the Milky Way's gravitational pull entirely. Studying them is crucial as they offer insights into the dynamics of galaxies, black hole interactions and the distribution of dark matter across the universe.

These stellar outcasts represent some of the most extreme objects in our Galaxy, with their trajectories providing a unique window into understanding galactic structure and evolution over cosmic timescales. The mechanisms that accelerate these stars to such tremendous velocities are among the most violent processes in the universe. Gravitational slingshot effects near the galactic centre, binary star disruptions and supernova explosions all contribute to creating these cosmic escapees. Their high speeds preserve information about ancient galactic events, making them invaluable archaeological records of our Galaxy's turbulent past.

A recent discovery revealed a 'super-Neptune' world orbiting such a fast moving, low-mass star. The system travels at an estimated 540 kilometres per second! If aligned with our solar system (with the star positioned where our Sun sits) the planet would orbit somewhere between Venus and Earth. Finding such objects is challenging. This system was first detected in 2011 through analysis of data from the Microlensing Observations in Astrophysics (MOA) survey conducted by the University of Canterbury in New Zealand.

Microlensing events occur when a mass passes between Earth and a distant object. When this happens, the intervening mass reveals its presence through changes in the background star's light curve. The New Zealand observations revealed signals from two celestial bodies, one approximately 2,300 times more massive than the other.

The star measured about 20% of the Sun's mass, while the planet weighed 29 times more than Earth. Alternatively, the microlensing data could indicate a closer planet about four times Jupiter's mass with an accompanying moon. To learn more, astronomers searched through data from the Keck Observatory and *Gaia* satellite. They located the star approximately 24,000 light years away, still within the Milky Way. By comparing the star's position in 2011 and 2021, they calculated its remarkable speed.

Travelling at 540 kilometres per second, it is unlikely to escape the Milky Way but astronomers are now searching for more examples. If they discover any moving faster than 600 kilometres per second, these systems will likely escape the Milky Way's gravity and enter intergalactic space millions of years from now.

High velocity stars captured by the Hubble Space Telescope. (NASA/ESA/JPL)

FACT 48

We are unlikely to be able to detect advanced civilisations by their industrial pollution

Given the amount of pollution we produce, it might be reasonable to assume that as we continue our search for advanced civilisations we should hunt for pollution signatures. After all, surely pollution is likely a key byproduct of an advanced, industrial civilisation? It seems perhaps not!

Over the decades, numerous methods for hunting advanced civilisations have been developed, from scanning stars for artificial radio signals or laser pulses to searching for evidence of water. These techniques have so far returned no positive results. Initiatives like SETI (search for extraterrestrial

ALH84001 is a meteorite from Mars that has naturally occurring polycyclic aromatic hydrocarbons inside it. (NASA)

intelligence) have used some of the world's most powerful radio telescopes to listen for signals. Meanwhile, the habitable zones of exoplanets have been probed for signs of water, which would suggest life may exist. To date, however, we remain in the dark about the existence of any alien neighbours.

Chemicals with the funky-sounding name polycyclic aromatic hydrocarbons (PAHs) have been suggested as one alternative way we might hunt down aliens. PAHs sprang into the headlines when they were detected inside a Martian meteorite, albeit naturally produced and not due to industrial pollution. Their discovery received significant attention since these chemicals are a known byproduct of life, and finding them buried in a Martian meteorite suggested some form of life may have existed at some point in Mars's history. The debate continues, but for a time it was believed that searching for hydrocarbons like these in planetary atmospheres could reveal alien life and even advanced civilisations.

PAHs do not, however, occur exclusively due to life's presence. There are other sources in space, such as the interstellar medium, but their origins are mostly associated with biological activity. A recent study focused on hydrocarbons that have measurable absorption cross-sections in the atmospheres of Earth-like exoplanets. An absorption cross-section measures the probability of an absorption process, such as particle scattering. The study chose four specific chemicals: naphthalene, anthracene, phenanthrene and pyrene which are hydrocarbons found in industrial pollution.

Drawing on evidence from Earth-based PAH concentrations, the study found that these levels have actually declined since the Industrial Revolution rather than increased. Learning from this, researchers ran simulations across various concentration ranges, hoping to prove the detection capabilities for identifying Earth-like civilisations.

The study also investigated whether current telescopes might limit detectability – after all, we need to know if these signatures are actually detectable with available technology. Their analysis relied on large telescope mirrors being able to resolve spectral details of the four target molecules. However, they found that telescopes with 6-metre, 8-metre or even 10-metre apertures would be unable to detect meaningful levels on alien worlds.

This represents an excellent example of research that does not yield a positive outcome, yet negative results in scientific research are equally valuable. Further research and laboratory-based measurements are now needed to improve the detectability of these molecules and perhaps help us find our first cosmic neighbour.

FACT 49

We have sent golden records out into deep space

Perhaps at risk of showing my age a little with this one, it is probably appropriate to explain first what a 'record' is! Back in the day, a record was a circular disc made of vinyl used to store and play back sound recordings, usually music. Upon its surface was a spiral groove that ran from the outer edge

The golden record on board the *Voyager* spacecraft. This gold aluminium cover was designed to protect the *Voyager 1* and *2* 'Sounds of Earth' gold-plated records from micrometeorite bombardment, but also served a double purpose in providing the finder with a key to playing the record. The explanatory diagram appears on both the inner and outer surfaces of the cover, as the outer diagram would be eroded over time. (NASA)

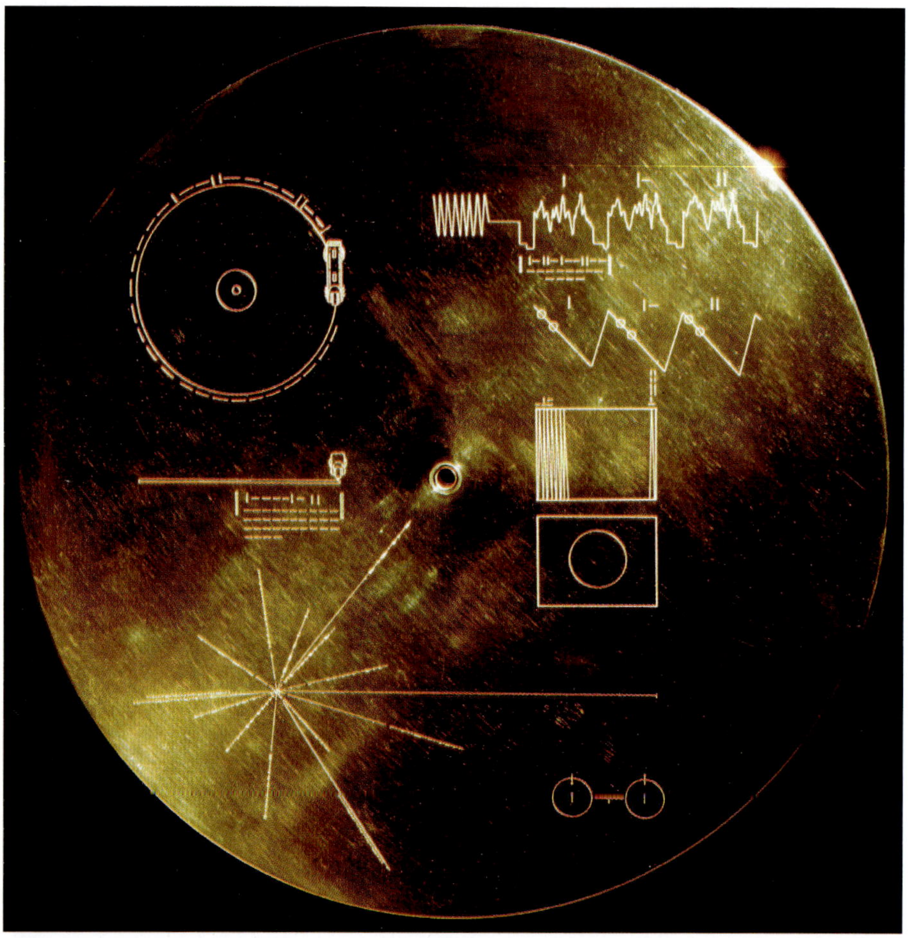

toward the centre, and within the groove were tiny variations that represented the waveform of the sound. These variations were analogue and directly corresponded to the shape of the sound waves. To play the record, the disc was spun at a specific speed while a needle sat within the groove. The movement of the needle up and down was converted to an electrical signal, amplified, and played through speakers.

Record technology was developed in the late 1800s and remained commonplace until the 1980s, when it began to decline in favour of digital formats. What has this got to do with space? The beauty of the record was that it was a physical device that typically travelled well and provided an excellent way of transferring sound information. The idea of using a record to send audio information to the stars was first discussed in 1976, when it was proposed to attach such an item to the outside of the *Voyager* spacecraft scheduled for launch later that decade.

Voyager 1 and *Voyager 2* were developed and launched by NASA in 1977 to explore the outer solar system. *Voyager 2* launched first, followed by *Voyager 1* just a few weeks later, with both undertaking fly-bys of the gas giants on a grand tour of the solar system. *Voyager 1* visited Jupiter and Saturn while *Voyager 2* visited all four outer planets: Jupiter, Saturn, Uranus and Neptune. Both spacecraft are now officially in interstellar space and still transmit information about the interstellar environment.

The *Voyager* spacecraft are now heading out in different directions: *Voyager 1* toward the constellation Ophiuchus and *Voyager 2* toward Telescopium. They will eventually run out of power, but even when they do, they both carry a golden record on board . These 30.48 cm records are gold-plated copper discs that contain messages from Earth to any civilisation that may one day find them.

They were created under the careful guidance of astronomer Carl Sagan and a team who curated a selection of sounds, music and greetings from Earth in 55 different languages. The idea was that they portray the diversity of life and culture on Earth, including music from Bach and Beethoven, along with pieces from other eras and cultures. There are also 116 photographs encoded on the record showing a variety of animals, humans, nature and scientific instruments. Much more than just a way to play music, these golden records have been designed as time capsules and a message to the universe, carrying greetings of peace and friendship to the stars.

FACT 50

Gold is created in deep space

It is not something often thought about – where do all the elements come from? Every atom inside you and me has at some point been synthesised deep inside a star. When the universe formed 13.8 billion years ago, it was composed largely of hydrogen with a tiny bit of helium. The process of nucleosynthesis inside stars was then responsible for turning these elements into the carbon, oxygen, neon, magnesium, silicon and iron we see today. Many other processes are responsible for creating other elements, but the heavier elements like gold and platinum are formed through a rather wonderful, if explosive, process in deep space.

Magnetars are a rare and extreme type of neutron star with incredibly strong magnetic fields, often up to a thousand trillion times stronger than Earth's. They form from the collapsed core of a massive star after a supernova explosion and are known for their intense bursts of X-rays and gamma rays. Occasionally, these magnetars undergo 'starquakes', releasing high-energy radiation through giant flares that can affect Earth's atmosphere even from hundreds or thousands of light years away. Studies have revealed that while supernova explosions create heavy elements, these magnetar flares could also be responsible for forming heavy elements by rapidly fusing neutrons into heavier atomic nuclei.

Elements like gold and platinum are primarily formed in explosive cosmic events such as supernovae and neutron star collisions, where intense energy creates the conditions for nuclear reactions. These processes scatter heavy elements throughout the universe, enriching the interstellar medium and providing the building blocks for planets and life. Studies have now shown that magnetars too contribute to the formation of these elements through their powerful flares, potentially creating as much as 10% of the heavy elements, like gold, in our Galaxy.

The role of magnetars highlights how extreme environments, beyond the classic supernova explosions, play a vital part in the cosmic alchemy that enriches the universe. Their unique magnetic fields and violent flares provide conditions unlike any other known process, forging elements at extreme levels.

As we continue to study these mysterious objects, we gain new insights into how the fundamental building blocks of life and civilisation have been created in the stars. Each precious metal and element on Earth carries a story that spans billions of years and countless cosmic events, connecting us intimately to the stars themselves. This cosmic connection reminds us that the matter making up our world is not static but part of an ongoing, dynamic cycle of creation and destruction across the universe.

Artist impression of a magentar. (ESO/L. Calçada)

FACT 51

Astronauts cannot cry properly because there's no gravity to pull tears away

Ever wonder what happens when you cry in space? It is actually pretty weird. Down here on Earth, we are so used to gravity doing its thing that we never really think about how it affects something as simple as shedding a few tears.

When you are floating around in zero gravity, crying becomes a somewhat unusual experience. Normally, your tears just roll down your face thanks to gravity, clearing your eyes so you can see properly. But in space? Forget about it. Instead of those familiar tear tracks, you end up with these perfect little bubbles of water just sitting on your eyeballs. They keep growing bigger and bigger until they finally break away and float off into the air like tiny water balloons.

Astronauts say it feels really strange. You know how a good cry can actually make you feel better? Well, that does not really work in space. The tears just

Astronaut Clayton Anderson observes a floating bubble of water on board the Space Shuttle *Discovery*. (NASA)

sit there stinging your eyes and making everything blurry. It is like your body wants to have this emotional release, but it gets stuck halfway through the process. It is a bit like wanting to yawn but you cannot!

Chris Hadfield, the famous Canadian astronaut who spent time on the International Space Station actually made a video showing this whole process; you can see the tear just growing into a big bubble around his eye instead of rolling away like it should. The video really drives home how different everything is up there, even the simple stuff we take for granted.

And here's the thing – it makes you realise that crying is not just about emotions. It is actually pretty physical too. Part of what makes us feel better after a good cry is the tears leaving our body. When they are just floating around your face, it does not have the same effect.

This might seem like a pretty minor thing, but it is actually a big deal when you consider long space missions. Imagine being stuck in a tiny spacecraft for months or even years, already feeling homesick and isolated, and then not even being able to have a proper cry when you need one.

As we start planning trips to Mars and other far-off places, we are going to need to deal with these rather more psychological issues that ultimately help us to manage our emotions. The focus is often on physical issues like muscle and bone density degradation but long-duration spaceflight brings other issues too. Sometimes it is the small, everyday human things that can really challenge us.

So next time you have a good cry, maybe take a second to appreciate that those tears are actually falling where they are supposed to. In the endless blackness of space crying may become more of a problem than you think!

FACT 52

Some stars have surface temperatures cooler than your morning coffee

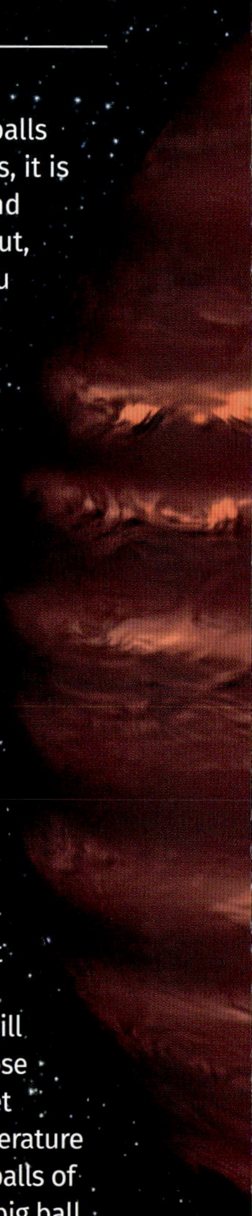

When you think of the stars, you probably picture them as blazing hot balls of fire that would instantly vaporise anything that got too close. And yes, it is true that in most cases stars are exactly that: our Sun clocks in at around 5,500 degrees on its visible surface, which is definitely not touchable. But, out in the depths of space, there are stars that are cool enough that you could theoretically give them a high-five without getting fried!

These cool stars are called brown dwarfs, and ultimately they are failed stars. They were never quite massive enough to kick-start the nuclear fusion process that makes regular stars so hot. Instead of raging nuclear fires, instead they are just sitting there, slowly cooling off like a giant space embers.

The coolest brown dwarfs that have been discovered have a surface temperature around room temperature, somewhere around 20°C, cooler than a hot summer day. Some are even chillier, dropping down to what you'd feel on a mild spring afternoon. It is amazing to think that there are objects floating in the freezing vacuum of space that are more comfortable to touch than your morning coffee ... or tea!

Infrared telescopes were used to spot these cool stellar cousins, since they do not give off much visible light. They are out there glowing faintly in infrared radiation, somewhat like space heaters, but not the sort of space heaters that warm your home, these are actual *space* heaters. The Spitzer Space Telescope found some of the first brown dwarfs and now we know there are probably hundreds if not thousands of them in our region of our Galaxy.

Of course, there is a catch, and It is a big one. Even though these stars will not burn you, they are still massive objects with serious gravity. Getting close enough to actually touch one would be like trying to hug a planet. You'd get crushed long before you could feel how surprisingly comfortable the temperature is. That's coupled with the challenge that brown dwarfs, like stars, are big balls of gas so you don't actually have a solid surface you can touch, instead it's a big ball of gas. At some point, light gets emitted and this is called the 'surface'.

But still, it is pretty cool to know that somewhere out there in the darkness between stars, there are brown dwarfs just drifting around at temperatures you could actually survive. They are like the universe's way of saying that not everything in space has to be an extreme, hostile environment.

Next time you are looking up at the night sky, remember that mixed in with all those blazing suns are some stars you could literally shake hands with, if only you could survive the approach and find the surface.

Artist's impression of a brown dwarf. (NASA/JPL Caltech)

FACT 53

There are more stars in the Milky Way than humans *ever* born!

Here's a number that will blow your mind: current studies suggest there are somewhere between 100 and 400 billion stars in our Galaxy alone. To put that in perspective, if you tried to count them all at one star per second, it would take you over 3,000 years just to reach 100 billion. And that's just our own Milky Way Galaxy.

Now let's talk about humans. How many people do you think have ever lived on Earth throughout all of human history? Take a guess, It is probably a lot less than you're thinking. Demographers estimate that roughly 117 billion humans have been born since our species first appeared around 300,000 years ago. That includes everyone from the first *Homo sapiens* wandering around Africa to your great great grandmother to the person born five minutes ago.

Even if we take the most conservative estimate of stars in the Milky Way, that low-end 100 billion figure, we are already pretty much tied with the total number of humans who have ever existed. But most astronomers think the real number is closer to 200-400 billion stars, which means our Galaxy has roughly two to four times more stars than people who have ever drawn breath on this planet.

Think about that for a second. Every single person who ever lived, every pharaoh, every inventor, every artist, every kid who ever blew out birthday candles and of course, every astronomer, all of humanity combined still does not add up to the number of suns burning in our Galaxy alone. And there are an estimated *2 trillion* galaxies in the observable universe. The numbers get so big they really seem to stop making sense.

It puts things in perspective. We spend so much time focused on our daily lives, our problems, our little corner of the world. But we are living in this massive cosmic city with hundreds of billions of stellar neighbours, most of which probably have their own planets orbiting around them.

The next time you're feeling overwhelmed by how crowded Earth seems with its 8 billion people, just remember that we are actually pretty sparse when you zoom out to the galactic level. There are more stars in the sky above you right now than all the humans who have ever looked up at them combined.

It is humbling and amazing at the same time. We are part of something so much bigger than we usually imagine, floating around in a galaxy that's absolutely packed with other suns, probably other worlds, maybe even other forms of life, wondering about the same cosmic maths that we are.

The multitude of stars in our Galaxy revealed in this image of the Scutum Star Cloud. (Hypatia Alexandria)

FACT 54

Only a fraction of stars are white

When you look up at the night sky, pretty much every star looks like a tiny white dot, right? That's what most people think stars look like, brilliant white points of light scattered across the darkness. But only a small fraction of stars are actually white. The universe is a lot more colourful than our eyes let us see.

The truth is, stars come in a whole rainbow of colours: red, orange, yellow, blue and, yes, some white ones too. Our Sun, for example, is yellow, not white. It is what we call a G-type star, and it has a yellowish glow when you see it through the right filters. But since we are used to seeing sunlight as 'normal' light, our brains kind of treat it as white.

Most stars in our Galaxy are actually red dwarfs – small, cool stars that glow with a dim reddish light. They make up about 75% of all the stars in the universe, but they are so faint that we can barely see them with the naked eye. That's why the night sky looks dominated by brighter, whiter stars. We are seeing the cosmic equivalent of the loudest voices in the room, not the most common ones.

The really hot, massive stars burn blue or blue/white, and they are absolutely gorgeous when you see them through a telescope. Stars like Rigel in the constellation Orion are brilliant blue giants that make our yellow Sun look pretty mellow by comparison. On the flip side, cool giant stars like Betelgeuse glow orange or red, giving off this warm, cozy light that's nothing like the harsh white we usually imagine.

So why do they all look white to us? It is partly because our eyes are not great at picking up subtle colours in dim light, and partly because Earth's atmosphere scatters and changes the light before it reaches us. Plus, when you're looking at something that's basically a distant point of light, your brain tends to default to seeing it as white.

If you want to see the real colours of stars, grab a pair of binoculars or a small telescope some time. You'll be amazed at how colourful the night sky actually is. There are golden stars, red giants and brilliant blue beacons scattered everywhere.

The scattered stars of the globular cluster NGC6355 clearly show the different colours of the stars.
(NASA/ESA/Hubble)

FACT 55

There are diamonds scattered across the galaxy

When you think of diamonds, you probably tend to visualise expensive jewellery or maybe even industrial cutting tools. But there are literally trillions of diamonds floating around in space, scattered throughout our Galaxy like cosmic glitter. And we are not talking about tiny specks, some of these space diamonds are the size of entire planets.

The most famous diamond in space is probably BPM 37093, a white dwarf star that astronomers have nicknamed 'Lucy' after the Beatles song 'Lucy in the Sky with Diamonds'. This stellar corpse is a massive diamond roughly the size of Earth, containing 10 billion trillion trillion carats of crystallised carbon. It is sitting about 50 light years away, slowly cooling down over billions of years.

But Lucy is not alone. Scientists estimate that most white dwarf stars, which are the leftover cores of dead stars like our Sun, eventually become giant diamonds as they cool. Since white dwarfs make up a significant chunk of the stars in our Galaxy, that means there are probably millions of planet-sized diamonds just hanging out in space, waiting to be discovered.

These cosmic gems form through an incredible process. When a star runs out of fuel, it sheds its outer layers and leaves behind a super-hot, dense core made mostly of carbon and oxygen. As this white dwarf cools over billions of years, the intense pressure and slowly dropping temperature causes the carbon to crystallise into diamond. It is like nature's ultimate pressure cooker, creating gems on a scale that makes Earth's diamond mines look like sandbox toys.

Scientists have even found evidence of diamonds in meteorites that have crashed on Earth. These are not formed by the same process as white dwarf diamonds, they are created when asteroids smash into each other at incredible speeds, generating the heat and pressure needed to turn carbon into diamond instantly. Some of these space diamonds are harder than anything we can make here on Earth.

There are also diamond 'clouds' floating around in space, formed when carbon-rich stars die and spew their contents into the cosmos. These microscopic diamond particles drift through the Galaxy, occasionally clumping together into dust clouds that sparkle with their own light.

But before you start planning any space mining expeditions, remember that getting to these cosmic treasures would be slightly more challenging than a trip to your local jewellery store. It is pretty amazing though to think that we are living in a Galaxy absolutely loaded with the universe's most precious gems, just waiting out there in the darkness.

Illustration of a white dwarf star composed of compressed carbon, a diamond star. (Sephirohq)

FACT 56

A planet has been found clinging to a star racing through space at over 500 kilometres per second

Imagine a star hurtling through space at over 500 kilometres per second – that's fast enough to travel from London to New York in just 11 seconds. Now picture an entire planet being dragged along for the ride. That's exactly what astronomers have discovered: the first planet ever found orbiting what's called a *hypervelocity star*.

These runaway stars are among the fastest moving objects in our Galaxy. They get kicked out of their home regions by powerful gravitational forces, often after getting too close to massive black holes or colliding with other stars. Some move so quickly they can eventually escape the Milky Way (or other galaxies) entirely, becoming wanderers in the vast emptiness between galaxies.

Finding this speeding star was not easy. The discovery began in 2011 when astronomers noticed something unusual. They were using a technique called *gravitational microlensing*, which works like a space-based magnifying glass. When a massive object passes between Earth and a distant star, it bends the light from that background star, creating a temporary brightening effect.

The observations revealed two objects: a star about one-fifth the mass of our Sun, and a planet roughly 29 times heavier than Earth, making it what scientists call a 'super Neptune'. If this system were placed in our solar system with the star where our Sun sits, the planet would orbit somewhere between Venus and Earth.

To figure out just how fast this system was moving, astronomers had to play a game of 'spot the difference'. They compared the star's position in 2011 with where it appeared 10 years later in 2021, using data from powerful telescopes and space missions. This allowed them to calculate that the entire system is racing through space at about 540 kilometres per second.

The star and its planet are currently about 24,000 light years away from us, still within our Galaxy but moving at such an incredible speed that they might eventually break free from the Milky Way's gravitational grip. If their speed increases to around 600 kilometres per second, they will be travelling fast enough to eventually become intergalactic travellers, journeying into the empty space between galaxies millions of years from now.

This discovery is significant because it helps us understand how galaxies work and how black holes can influence entire star systems. These high-speed escapees also provide clues about dark matter, the mysterious invisible substance that makes up most of the universe's mass. As astronomers continue to study this remarkable system, they are hoping to learn more about both the planet and its breakneck journey through the cosmos.

Artist impression of a super-Neptune. (Pablo Carlos Budassi)

FACT 57

Jumping on deep space trips helps stop muscles from wasting away

When astronauts venture into space, their bodies face a silent enemy: weightlessness. Astronauts in orbit experience this as their spacecraft is constantly falling towards the Earth but the curve of the Earth is falling away from them. The spacecraft, and anything inside it is in a constant state of falling without hitting the ground. That's what gives astronauts the experience of floating around.

Without the usual gravity experience, bones become weaker and muscles begin to waste away. This is especially problematic for muscles in the legs and

Astronaut Sunita Williams running on board the International Space Station to maintain bone and muscle density. (NASA)

back, which normally work hard to keep us upright against gravity's force. The longer astronauts stay in space, the more their bodies deteriorate, making it difficult to walk and move normally when they return to Earth.

This challenge becomes even more critical as we plan missions to Mars (or further into space), which would require a nine-month journey each way. Future lunar bases will also test human endurance over extended periods since, while astronauts will not experience full weightlessness on such bases, the gravitational force on their bodies will be far less than they are used to on Earth. Space agencies are constantly searching for better ways to keep astronauts healthy during these long missions.

Surprisingly it has recently been discovered that jumping might be one of the best exercises for protecting our joints in space. Scientists studied how jumping affects cartilage, the smooth, cushioning tissue that sits between our bones and allows our joints to move without pain or grinding. Unlike many other body tissues, cartilage heals very slowly, making it especially important to protect. Extended periods of inactivity, whether from bed rest or spaceflight, can cause cartilage to break down. Space radiation can also make this deterioration worse.

Currently, astronauts spend up to two hours daily running on specially designed treadmills to maintain their health. While this helps slow cartilage breakdown, jumping exercises could be even more effective. Exactly how is this is possible when you are completely weightless? Astronauts can use treadmills and now jumping, by employing the elastic properties of great big rubber bands and springs.

This has been tested on mice! Researchers divided mice into different groups over nine weeks. Mice that had restricted movement experienced a 14% reduction in cartilage thickness in their joints, similar to what happens to astronauts in space. However, mice that performed jumping exercises just three times a week showed remarkable improvements. Their cartilage was 26% thicker than mice that did not exercise at all.

The benefits did not stop there. The jumping exercises also strengthened bones, with the active mice showing 15% higher bone density than the control group. While more research is needed to confirm these benefits apply to humans, the results are promising. If jumping proves equally effective for people, it could become a standard part of astronaut training both before and during missions.

FACT 58

X-ray emissions reveal a planet has been destroyed by a white dwarf

The Helix Nebula, also known as the 'Eye of God' for its striking resemblance to a giant eye staring back at us from space, is one of the most photographed objects in the night sky. Located 650 light years away in the constellation Aquarius, this colourful cloud has captivated amateur and professional astronomers alike for decades.

This beautiful nebula tells the story of a star's death. About 10,000 years ago, a star similar to our Sun reached the end of its life. For billions of years, this star had maintained a delicate balance nuclear fusion in its core pushed outward while gravity pulled inward. Eventually, the star's gravity began losing the battle, causing its outer layers to be blown out into space, appearing as giant smoke rings. What remained was an incredibly dense, hot core called a white dwarf, no bigger than Earth but containing most of the original star's mass.

The expelled gas glows brilliantly as it is lit up by intense ultraviolet radiation from the white dwarf at the centre, creating the stunning colours we see today. The emissions are not just ultraviolet though. For over 40 years, astronomers have detected mysterious X-rays coming from the nebula too. This puzzled scientists because white dwarfs are not typically strong X-ray sources. Using powerful space telescopes, researchers have finally solved this mystery, and the answer is more dramatic than anyone expected.

The X-rays appear to be evidence of a planetary massacre. Scientists believe they are witnessing the aftermath of a Jupiter-sized planet that was torn apart by the white dwarf's immense gravitational forces. The doomed planet likely began its existence in the outer reaches of its solar system, similar to Jupiter in our own system. Over millions of years, gravitational interactions with other planets gradually pushed it closer and closer to its star. As the planet spiralled inward, it crossed a deadly threshold called the *Roche limit*, the point where a white dwarf's gravity becomes so intense that it overcomes the planet's own gravity. The planet was stretched and torn apart and its remains formed a ring of debris that now orbits the white dwarf every 2.9 hours.

This planetary debris continues to fall onto the white dwarf's surface, heating up to millions of degrees and producing the X-rays that first alerted astronomers to the phenomenon. What makes this discovery particularly compelling is the consistency of the X-ray emissions. Observations spanning from 1992 to 2002 show the same steady brightness, suggesting an ongoing process rather than a one-time event. The regular 2.9-hour cycle in the X-ray signal provides strong evidence that we are seeing the final remains of a world that once may have had its own moons and weather systems.

It offers a sobering glimpse into the future – in about 5 billion years, our own Sun will undergo a similar transformation, potentially destroying the inner planets and perhaps the outer planets in the same process.

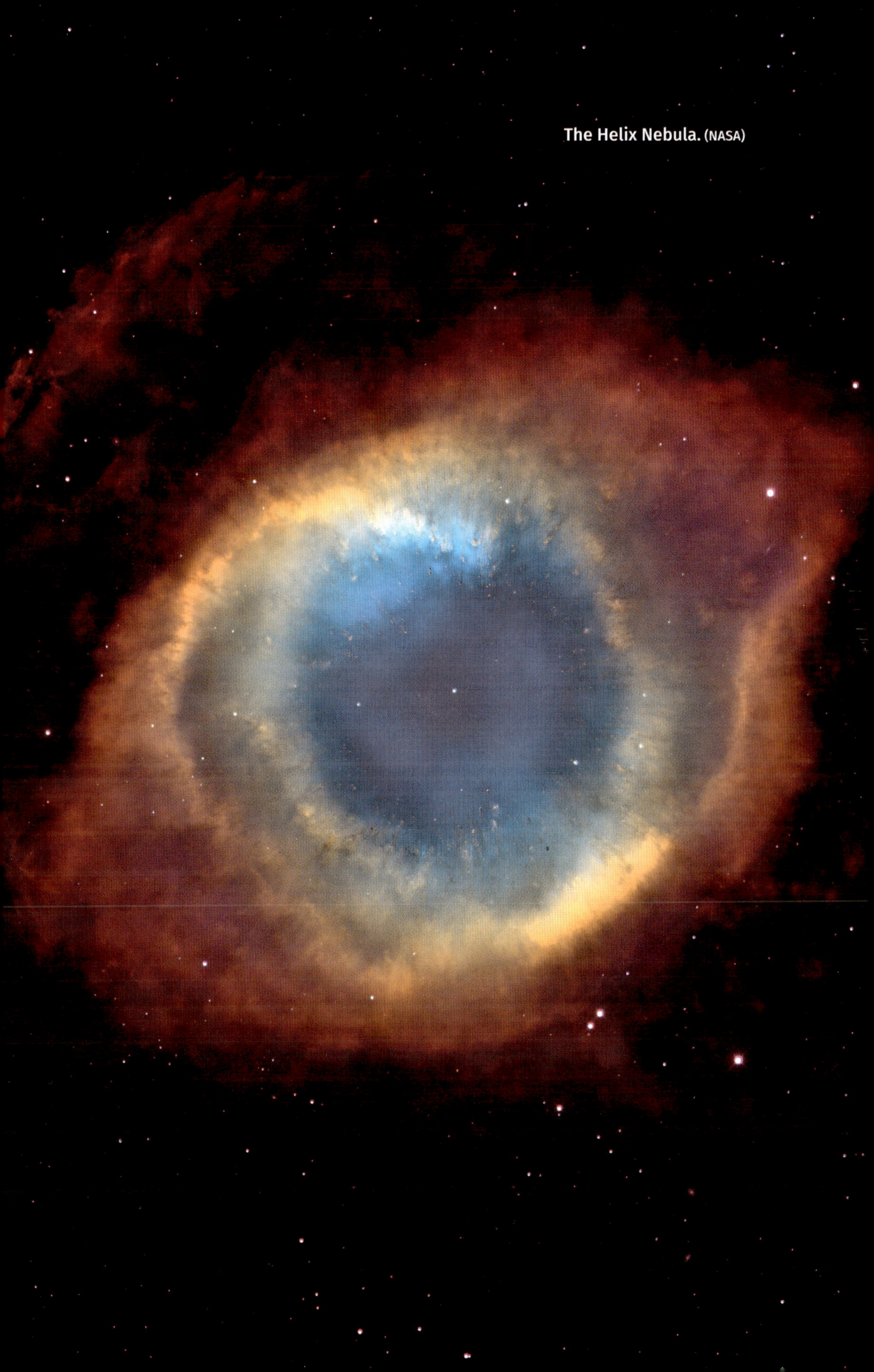

The Helix Nebula. (NASA)

FACT 59

You age slower when travelling faster

Here's one of my favourite facts in the book: the faster you move, the slower you age. It sounds like science fiction, but it is real, and it is happening to you right now, even if you cannot feel it.

It is called *time dilation*, and it is one of Einstein's most famous discoveries. Basically, time is not the constant, steady thing we think it is. Instead, it is more like a stretchy rubber band that gets pulled and warped depending on how fast you are moving through space.

When you are travelling around in a car or flying in a plane, you are actually ageing a tiny bit slower than someone standing still on the ground. We are talking incredibly small differences though, maybe a few billionths of a

Twins Mark and Scott Kelly, who undertook the famous 'twins experiment'. Scott went into space while Mark stayed on Earth. (NASA)

second over your entire lifetime. But the effect is real and measurable with super-precise atomic clocks.

The faster you go, the more dramatic this gets. If you could somehow travel at 90% the speed of light for a year, you'd come back to find that over two years had passed on Earth. Your friends would be two years older while you'd only aged one year. It is like having a real-life time machine that only goes forward.

This is not just theoretical. Astronauts on the International Space Station are moving about 17,500 miles per hour relative to Earth, so they are constantly experiencing this effect. After spending six months in space, an astronaut returns having aged about 0.01 seconds less than they would have on Earth. It is not much, but it is measurable.

The really crazy part is that from your perspective, time feels completely normal. If you were on that hypothetical 90% light-speed journey, your watch would tick normally, you'd age at the usual rate, and everything would seem fine. It is only when you compare your experience to someone moving at a different speed that the weirdness shows up.

GPS satellites deal with this all the time. They are moving fast enough that their onboard clocks run slightly slow compared to Earth clocks. Engineers have to constantly adjust for this time dilation effect, or your GPS would be off by miles within hours.

So next time you're rushing to catch a bus or sprinting for the train, remember you are not just trying to save time, you are literally slowing it down. Sure, the effect is so small you'll never notice it, but you are still participating in one of the universe's strangest tricks. Time, like space, is relative, and speed is your ticket to the cosmos's weirdest loophole.

FACT 60

Exocomets exist around distant stars

Just as our own solar system is filled with comets, we'd expect to find similar icy wanderers orbiting other stars. These 'exocomets' are incredibly difficult to spot directly due to their size, but astronomers have found a clever way to detect them by looking for the debris they leave behind.

Comets are essentially dirty snowballs made of ice, rock and dust. As they travel around a star or if they collide with each other or break apart, they create clouds of tiny pebbles and dust particles. It is these cosmic breadcrumbs that tell us where exocomets are hiding in distant star systems.

Using two of the world's most advanced radio telescope arrays, one in Chile's Atacama Desert and another in Hawaii, astronomers have been mapping these distant comet populations. These telescopes can detect the faint radio waves given off when comet debris collides and heats up.

The results have been pretty spectacular. Scientists have now captured detailed images of comet belts around 74 different star systems, revealing a surprising variety of structures. Some look like narrow rings, similar to our own Kuiper belt beyond Neptune. Others form wide discs that stretch across vast distances.

The shapes of these comet belts tell fascinating stories. Some are perfectly circular, while others are stretched into oval patterns. These distorted shapes often indicate the presence of unseen planets whose gravity is stirring up the comet populations, much as Jupiter and Saturn influence our own solar system's structure.

These exocomet belts exist in the frozen outer reaches of their star systems, typically located tens to hundreds of times farther from their stars than Earth is from the Sun. At such immense distances, temperatures plummet to between-150°C and -250°C, cold enough to freeze any water and preserve these ancient icy bodies for billions of years.

One of the most intriguing aspects of the discovery is that older star systems seem to have fewer comet debris particles. This makes perfect sense: after billions of years, most comets have either been destroyed in collisions or ejected from the system entirely. It is like watching an hourglass run out of sand. The research also shows that comet belts closer to their stars lose material faster,

Illustration of exocomets and various planet formation processes around Beta Pictoris.
(NASA/FUSE/Lynette Cook)

presumably because the warmer environment causes more rapid destruction of icy bodies.

While much of the recent focus in astronomy has been on discovering exoplanets, this research opens up an entirely new field of study. By understanding how comet populations evolve around other stars, we gain valuable insights into how planetary systems form and change over time, including clues about the early history of our own region of space.

FACT 61

Astronomers still do not know why more galaxies spin one way than the other

Galaxies are among the most magnificent structures in the universe. These vast collections of billions of stars, along with gas, dust and mysterious dark matter, are held together by gravity and come in beautiful shapes from spirals like our own Milky Way to smooth ellipticals and chaotic irregulars. They cluster together across space, forming a great web that stretches throughout the observable universe.

The James Webb Space Telescope (JWST), launched in 2021, has already revolutionised our understanding of the universe. Now it has uncovered something that has astronomers scratching their heads: galaxies do not seem to spin randomly as we'd expect.

When scientists analysed images of 263 distant galaxies whose images were clear enough to determine their rotation direction, they found something startling. About two thirds of these galaxies rotate clockwise, while only one third spin counterclockwise. In a truly random universe, we'd expect these numbers to be roughly equal, somewhat like flipping a coin and getting heads or tails.

The pattern is so obvious that anyone looking at the images can see it with their own eyes. It is a discovery that challenges our basic assumptions about how the universe works. The team of astronomers exploring this discovery have come up with two main theories for why this might be happening, and both would change our understanding of galaxies.

The first possibility is mind-bending – and I mean seriously mind-bending! Perhaps the entire universe was born spinning. Some theories, including the idea that our universe exists inside a massive black hole, actually predict this could happen. If true, it would require scientists to completely rethink current models of how the universe formed and evolved.

The second explanation is more down to earth. As our planet orbits around the centre of the Milky Way, we are constantly moving through space. This motion might create a kind of optical illusion making galaxies that spin opposite to Earth's direction appear brighter and easier to detect. In other words, we might be seeing more clockwise-spinning galaxies simply because they are easier to spot from our moving vantage point.

If Earth's movement is creating this observational bias, it would mean we need to recalibrate how we measure distances to far-off galaxies. This recalibration could potentially solve other outstanding puzzles that have been bothering astronomers for decades, including disagreements about how fast the universe is expanding and why some galaxies appear older than the universe itself.

Whether the answer lies in a spinning universe or observational bias, this discovery reminds us that there are plenty of secrets waiting to be uncovered.

Galaxies like M101 all seem to spin in one direction! (ESA/Hubble)

FACT 62

Exoplanets can be discovered by listening for radio waves

Finding planets beyond our solar system has become one of astronomy's greatest achievements and, to date, over 5,200 exoplanets have been discovered. Scientists have developed clever techniques to spot these distant worlds, even though they are invisible to our eyes. Some methods watch for the tiny dimming when a planet crosses in front of its star while others detect the subtle wobble a planet causes in its star's motion.

Radio telescopes like the Lovwell Telescope at Jodrell Bank can be used to listen for exoplanets. (Mike Peel)

Now there is an entirely new approach that sounds like something from science fiction: hunting for planets by listening to their radio signals. It is based on the fact that many planets, including Earth, are surrounded by invisible magnetic fields that act like protective shields against harmful space radiation. While we can detect the magnetic fields of stars using special techniques that analyse how light becomes polarised, planetary magnetic fields are incredibly weak and nearly impossible to measure directly from such great distances.

The problem is like trying to hear a whisper next to someone shouting – the star's powerful signals completely overwhelm any faint magnetic signatures from orbiting planets. Rather than trying to detect planetary magnetism directly, astronomers have turned to an indirect approach: listening for radio waves produced by planetary auroras. These light shows we are familiar with on Earth occur when energetic particles from the star interact with a planet's magnetic field, creating not just beautiful displays of light but also radio emissions.

This phenomenon happens through something called the cyclotron maser instability (ECMI) mechanism, a process where electrons spiralling around magnetic field lines produce specific types of radiation. It is essentially the same process that creates the spectacular auroras we see on Earth, except these also broadcast radio signals into space.

Radio emissions from auroras were first discovered coming from Jupiter in 1955, producing signals ranging from a few thousand to 40 million hertz. Each planet with a magnetic field creates its own unique radio fingerprint, providing valuable clues about the strength and structure of its magnetic environment.

However, detecting these signals does still present significant challenges. The radio emissions are directional, meaning we can only hear them if we are positioned in the right spot relative to the planet, a bit like trying to eavesdrop on a conversation: from the wrong angle we can hear nothing clearly.

To explore whether this technique could work for distant planets, astronomers ran computer simulations based on Jupiter and its largest moons. The results were encouraging, showing that with enough observational time and the right equipment, it should be possible to detect radio emissions from exoplanets.

The key to success lies in patience and advanced technology. Future radio telescopes will be far more sensitive than current instruments, making it possible to pick up these faint whispers from planets many light years away. This technique could reveal not just the presence of distant worlds, but also whether they have the magnetic fields necessary to potentially support life.

FACT 63

There are pumpkin-shaped stars

When astronomers look up at the night sky, they occasionally stumble across objects so strange that they cannot help but give them quirky nicknames. Case in point: there are actually multiple stars out there that have earned the title 'pumpkin star', and each one is bizarre in its own spectacular way.

The most famous pumpkin-shaped star is VFTS 102, a blue supergiant located about 160,000 light years away in the Large Magellanic Cloud (LMC). This cosmic oddball has earned its Halloween nickname by spinning itself into a pumpkin shape with a rotation speed of roughly 300 kilometres per second at its equator, that's over 670,000 miles per hour.

VFTS 102 is flattening itself from a perfect sphere into a squashed, bulging mess because of its incredible spin rate. The centrifugal force is so intense that it stretches the star outward at its equator while crushing it down at the poles, creating that distinctive pumpkin-shaped bulge. This monster weighs about 25 times more than our Sun and is spinning right on the edge of tearing itself apart. Scientists think it got this crazy rotation from a cosmic break-up – it was probably once dancing with another massive star until its partner exploded as a supernova, giving VFTS 102 a gravitational kick that sent it spinning like a top.

But VFTS 102 is not the only star to earn pumpkin status. There's also EPIC 211779126, a young star in the famous Pleiades cluster that's gained its name for completely different reasons. This stellar youngster has been nicknamed the 'pumpkin star' because of its weird dimming behaviour and unusual brightness variations that seem to pulse in mysterious patterns.

EPIC 211779126 is much closer to home than VFTS 102, only about 400 light years away, and it is giving scientists a front-row seat to study how young stars behave during their teenage years. While it might not be physically pumpkin-shaped like its spinning cousin, its erratic light patterns have made it just as intriguing to researchers trying to understand stellar evolution.

What's remarkable is that in a universe containing hundreds of billions of stars, we've managed to find multiple objects weird enough to remind us of Halloween decorations. Whether it is VFTS 102 spinning itself into a pumpkin shape or EPIC 211779126 flickering like a celestial candle, these pumpkin stars prove that the cosmos has a sense of humour and that even the most serious astronomical objects can sometimes look absolutely ridiculous.

Who knows how many more pumpkin-shaped oddities are hiding out there in the darkness, waiting to surprise us?

Location of VFTS 102 in the LMC. (ESO)

FACT 64

The universe has no centre

No matter where you are in the cosmos, Earth, Alpha Centauri or some random galaxy billions of light years away, you'll always appear to be at the centre of everything. It is like being in the ultimate hall of mirrors.

It is actually one of the most profound insights in modern astronomy. When we look out into space, we see galaxies racing away from us in all directions, and the farther away they are, the faster they are moving. At first glance, this makes it seem like we are sitting at the centre of some massive cosmic explosion, with everything fleeing from our particular patch of the universe.

But here's the kicker: every other observer in the universe, no matter where they exist, would see exactly the same thing. An alien astronomer in the

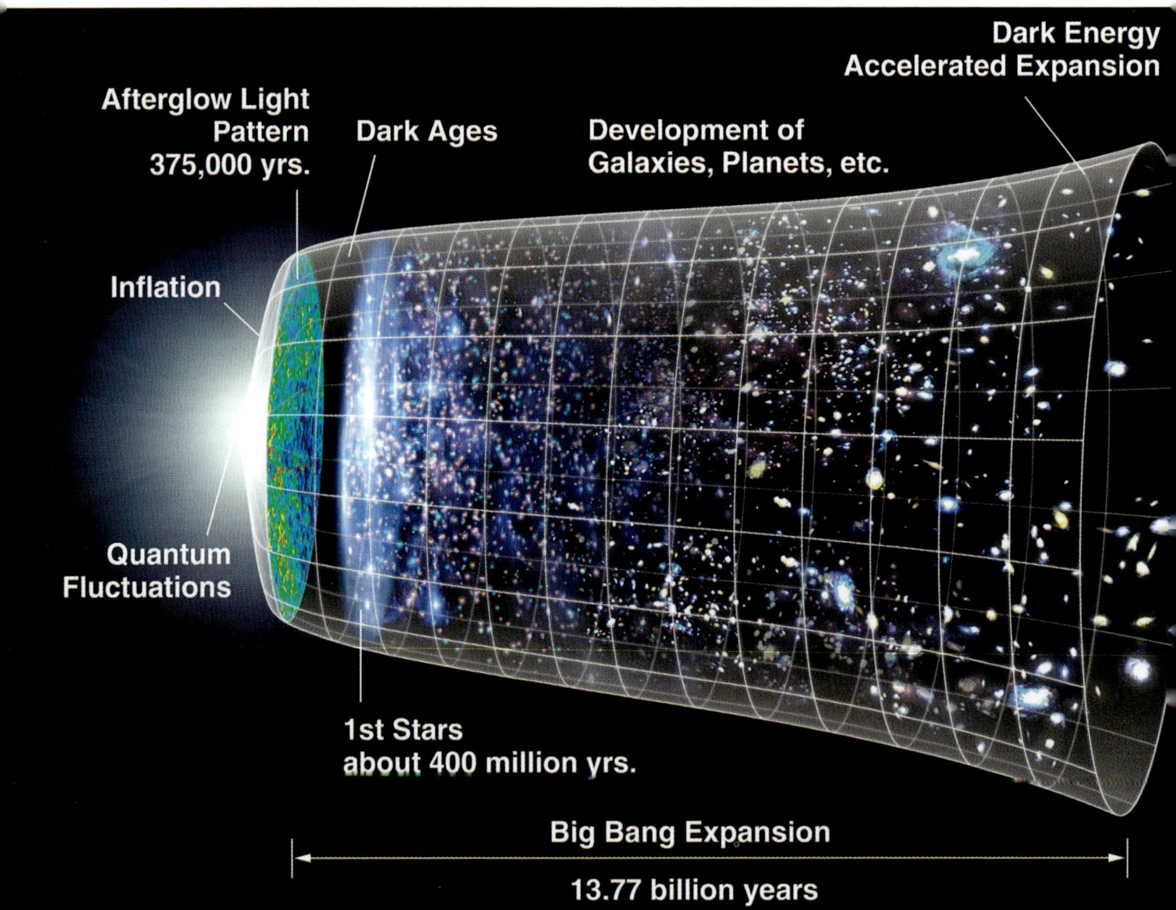

Andromeda Galaxy would look around and see all the galaxies rushing away from them too. A scientist in a galaxy 10 billion light years away would observe the same pattern. Everyone thinks they are at the centre because, in a very real sense, everyone is.

The secret lies in understanding how space itself works. The universe is expanding like a balloon, stretching outwards, with the galaxies painted on its surface, in one of which we live. As the balloon expands, every point on the surface sees every other point moving away from it. There's no special centre point on the balloon's surface, just like there's no centre to the universe.

This mind-bending reality comes from the Big Bang, but not in the way most people imagine it. The Big Bang was not an explosion that happened at a specific point in space, it was *space itself coming into existence*, everywhere at once. It is like asking what's north of the North Pole. The question does not really make sense because 'north' loses its meaning at that point.

Einstein's theory of relativity predicted this centreless universe long before we had the observations to prove it. Space and time are woven together into what physicists call space-time, and this fabric has been stretching ever since the universe began. The expansion is not happening *in* space, it is space itself expanding.

This means that if you could somehow step outside our universe and look at it from the outside, you would not see a sphere expanding into empty space. You'd see ... well, nobody knows what you'd see, because 'outside the universe' might not even be a meaningful concept.

So the next time you're feeling small and insignificant in the vast cosmos, remember this: from the universe's perspective, you really are at the centre of everything. You're just sharing that honour with every other point in existence.

A graphical representation of the expansion of the universe from the Big Bang to the present day. (NASA/WMAP Science Team)

FACT 65

Black holes can evaporate

Nothing can escape a black hole, right? Well, that's what everyone thought until Stephen Hawking came along and dropped one of the biggest bombshells in physics: black holes actually evaporate. They slowly leak energy and shrink over time until they completely disappear in a final flash of radiation.

At the edge of a black hole's 'event horizon' (the boundary beyond which nothing can escape once it passes), quantum mechanics gets really strange. Empty space is not actually empty, it is constantly bubbling with pairs of particles and antiparticles that pop into existence and immediately annihilate each other. Usually, this happens so fast that we never notice it. But near a black hole's event horizon, something unexpected happens.

Sometimes one particle from these quantum pairs gets sucked into the black hole while its partner escapes into space. The escaping particle becomes real radiation that carries away a tiny bit of the black hole's energy. Since energy and mass are equivalent, this means the black hole is losing mass bit by bit. It is like the universe's slowest leak, but it is constantly happening.

Unexpectedly, smaller black holes evaporate faster than big ones. A black hole with the mass of our Sun would take a ridiculous amount of time to evaporate, longer than 10^{67} years, which makes the current age of the universe (a mere 13.8 billion years) look like a blink of an eye. But a tiny black hole the size of a proton would vanish in a fraction of a second.

This creates what scientists call the *information paradox*. If a black hole completely evaporates, what happens to all the information about the stuff that fell into it? Does it just disappear forever? This question has been driving physicists crazy for decades because it seems to violate fundamental laws about how information works in the universe.

As a black hole gets smaller through evaporation, it actually gets hotter and starts radiating faster. The final moments would be spectacular, the black hole would heat up to trillions of degrees and explode with the energy of millions of hydrogen bombs. Of course, this will not happen to any black holes we know about for an incomprehensibly long time.

The discovery of *Hawking radiation*, as this evaporation process is called, connects three of the biggest theories in physics: general relativity, quantum mechanics and thermodynamics. It shows that even the most extreme objects in the universe are not permanent. Given enough time, everything, even black holes , will fade away into the dark.

Image of the black hole at the centre of our Milk Way Galaxy captured by the Event Horizon Telescope. (EHT Collaboration)

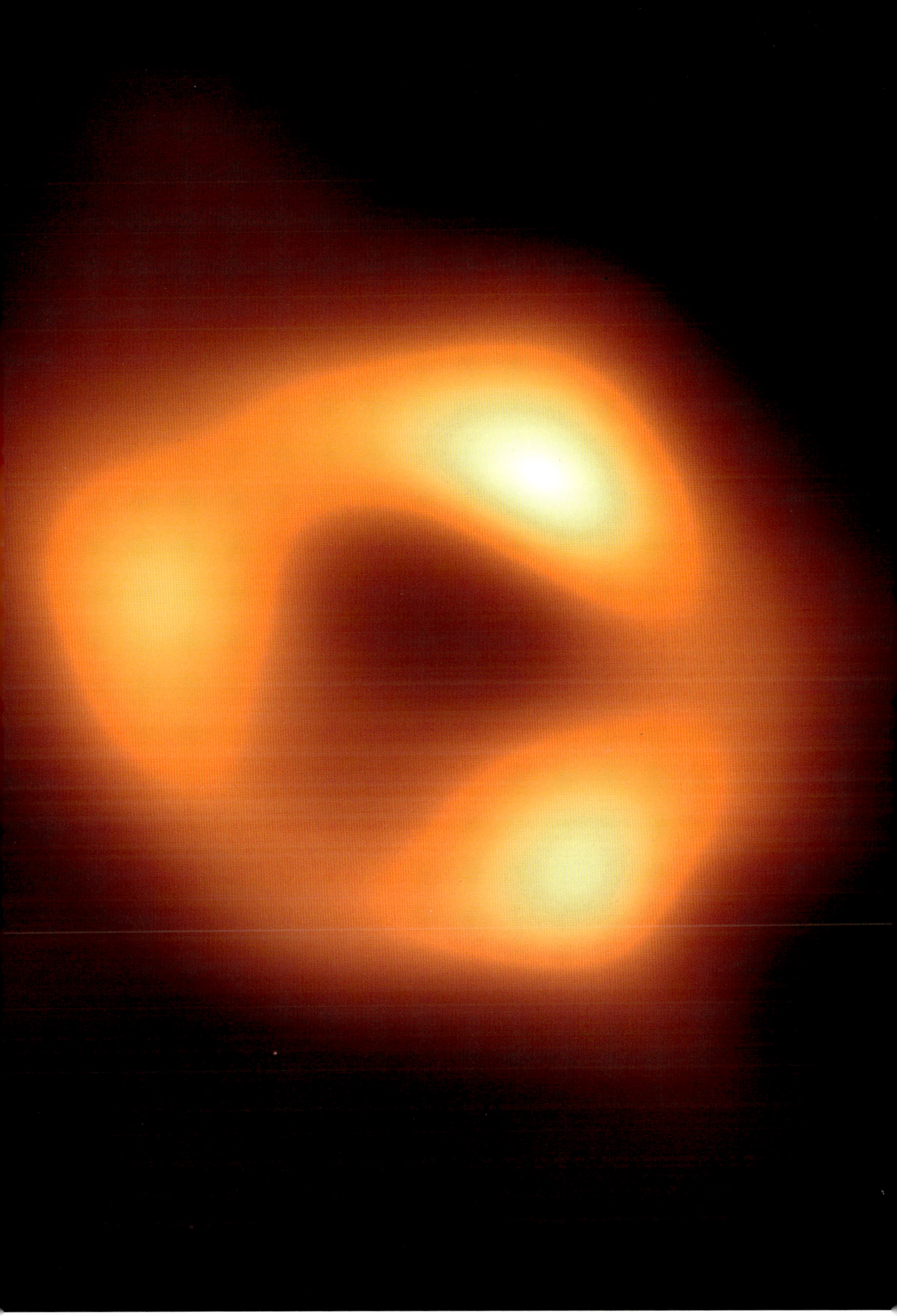

FACT 66

The solar system is moving through space at 828,000 kilometres per hour

Right now, as you sit reading this book, you are hurtling through space at an incredible 828,000 kilometres per hour. That's fast enough to circle the entire Earth in just three minutes, yet you cannot feel even the slightest sensation of movement. This mind-boggling speed represents how fast our entire solar system is racing through space.

The reason we do not notice this breakneck pace is the same reason passengers on a smooth airplane flight cannot feel they are traveling at 900 kilometres per hour – when everything around you moves at the same speed, there's no sensation of motion. Our planet, the Sun, and all the other planets, are locked together in this journey, swept along as one unified system through the galaxy.

Our solar system's journey through space is actually the result of several different movements happening simultaneously. Earth orbits the Sun at about 107,000 kilometres per hour, while the Sun itself orbits around the centre of the Milky Way Galaxy at roughly 720,000 kilometres per hour. But the Galaxy is not standing still either. It is moving through space toward neighbouring galaxies.

Illustration of the bodies in our solar system. **(CactiStaccingCrane)**

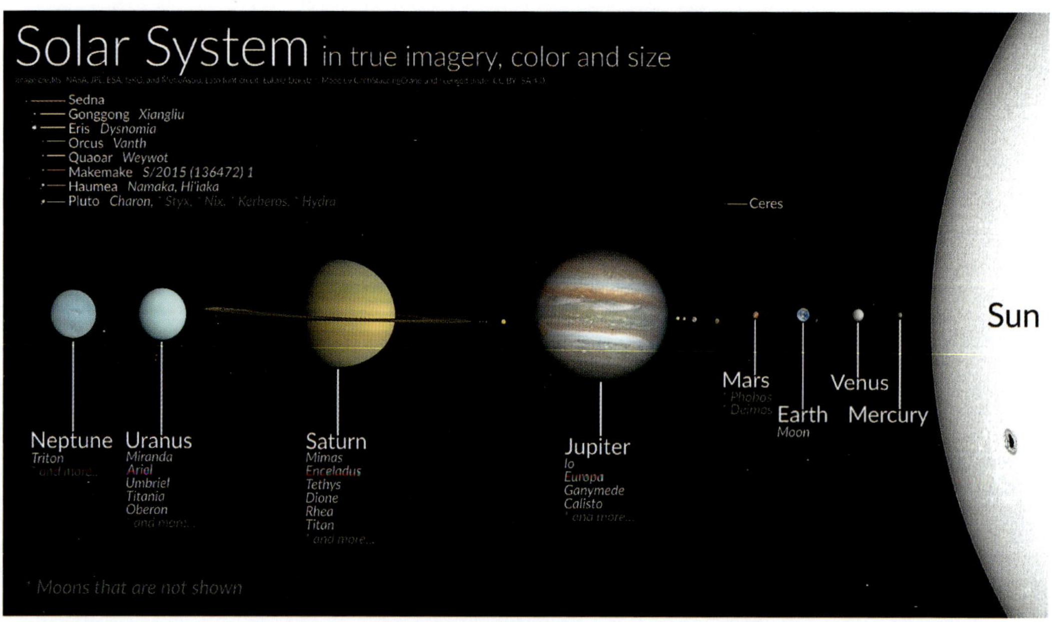

When astronomers add up all these different motions, they calculate that our solar system is moving at approximately 828,000 kilometres per hour relative to the cosmic microwave background radiation – the leftover glow from the Big Bang that provides a kind of universal reference frame for measuring motion.

Moving at this incredible speed, our solar system completes one orbit around the Milky Way's centre roughly every 225 million years, a period called a *galactic year*. During Earth's entire history, our planet has made this journey about 20 times. Dinosaurs lived when Earth was in a completely different part of the Galaxy than where we find ourselves today.

The Milky Way itself is on a collision course with the nearby Andromeda Galaxy, approaching at about 400,000 kilometres per hour. Do not worry though – this galactic merger will not happen for another 4.5 billion years, and when it does occur, the vast distances between stars mean our solar system, in whatever state it is in by then, will likely remain intact.

Scientists determine our speed by studying the cosmic microwave background (CMB). This radiation has been traveling through space for nearly 14 billion years. By measuring how this radiation appears slightly warmer in the direction we are moving and cooler in the direction we are moving away from, astronomers can calculate our motion through the universe with remarkable precision.

This journey we are on puts our place in the universe into perspective. We are not just residents of Earth or even citizens of the solar system – we are passengers on a cosmic voyage that has been underway for billions of years. Every sunset you've ever watched, every star you've ever wished upon, happened while racing through space faster than any human made object has ever travelled. Our cosmic address is constantly changing, yet home feels perfectly still.

FACT 67

The universe will eventually become cold, dark and empty

I am sorry to say that the universe has a rather dark destiny. According to our best understanding of physics and cosmology, the universe that is currently home to billions of galaxies will eventually become a cold, dark and almost empty expanse. This inevitable fate, known as the *heat death* of the universe, represents the ultimate triumph of entropy (or disorder.)

Before we think about the fate of the universe, let's take a brief look at the universe we live in today. It is a busy place. Stars burn brightly, galaxies spiral

Image of Sirius and Sirius B which is the closest white dwarf to Earth. Eventually the universe will be full of slowly cooling white dwarf stars. (NASA/ESA)

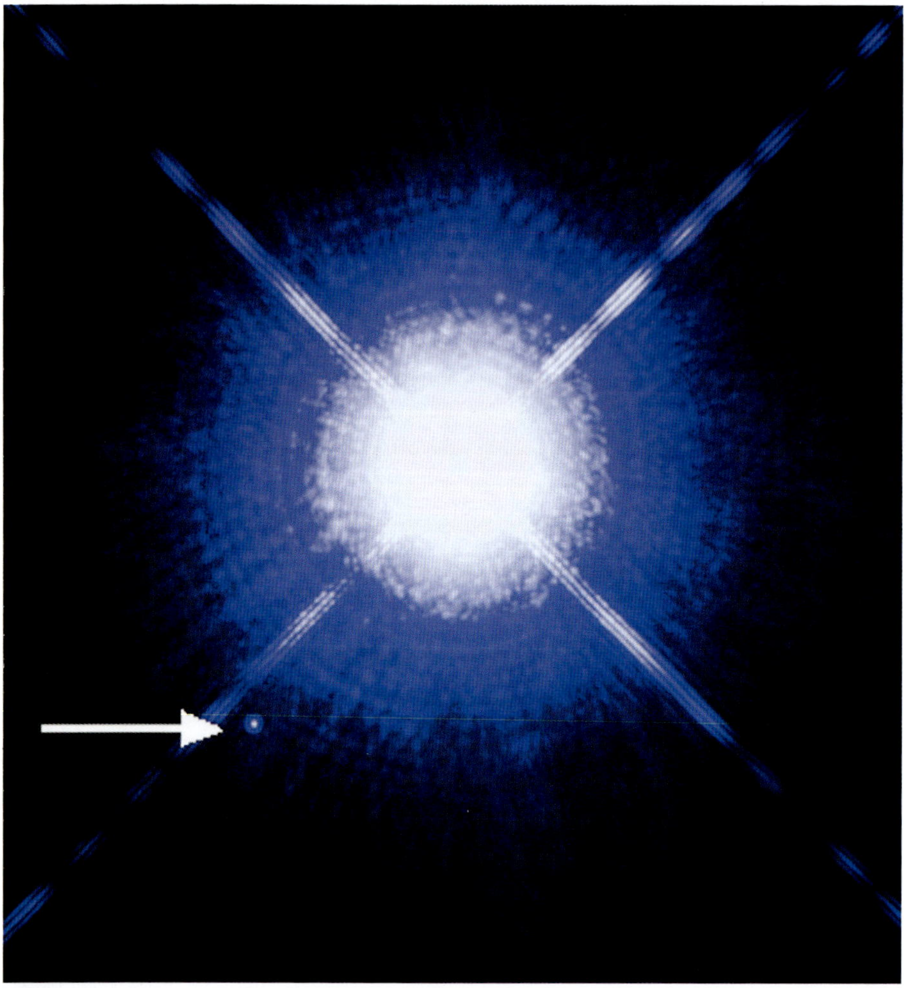

rather majestically and planets orbit in ways we have understood for at least the last 400 years. But this activity comes at a cost. Every fusion reaction in a star's core, every collision between particles, every photon travelling through space represents energy being transformed and dispersed. The second law of thermodynamics tells us that entropy – essentially, disorder – always increases in an isolated system, and the universe is the ultimate isolated system.

The timeline toward this grim future unfolds over unimaginably vast timescales. In roughly 5 billion years, our Sun will exhaust its hydrogen fuel and expand into a red giant before eventually becoming a white dwarf star. Similar fates await all stars, though the most massive ones will explode as supernovae, leaving behind neutron stars or black holes. The smallest red dwarf stars, however, will continue burning for trillions of years, far longer than the universe has currently existed.

The expanding and cooling universe means that the gas needed to create new stars becomes so diluted that new star formation gradually ceases. Eventually, around 100 trillion years from now, the last star will flicker out, leaving the cosmos illuminated only by the faint glow of white dwarf stars slowly cooling over hundreds of trillions of years.

But even this twilight phase will not last forever. Protons, the building blocks of atoms, may themselves decay over periods so vast they defy human comprehension, perhaps 10 undecillion years, that's 10 followed by 34 zeros! If this happens, all ordinary matter will eventually disintegrate into subatomic particles and radiation. Even the most stable structures, like planets and dead stars, will crumble into elementary particles.

The only remaining massive objects will be black holes, which will become the universe's longest-lived inhabitants. However, due to Hawking radiation, even these will eventually evaporate. Supermassive black holes, the giants lurking at galactic centres, will take an almost eternal 10 googol years (10 with 100 zeros) to completely disappear.

In this final epoch, dubbed the 'dark era', the universe will be nearly empty, containing only elementary particles, photons and neutrinos scattered across vast distances of empty space. The temperature will hover just above absolute zero, and the remaining particles will be so spread out that meaningful interactions become impossibly rare.

While this endgame lies incomprehensibly far in the future, it reminds us of the preciousness and impermanence of our current moment in the history of the cosmos. We exist during a brief, brilliant epoch when the universe is alive with stars, planets and, on at least one world we know of, life itself contemplating its own fate.

FACT 68

There is a galaxy with no stars

Imagine a galaxy so dark that it is almost invisible, containing virtually no stars despite being nearly as massive as our entire Milky Way. That's a concept I struggled to get my head round, since galaxies are mostly made up of stars! This is not science fiction though: astronomers have discovered galaxies that are essentially 'galactic ghosts', challenging everything we thought we knew about how galaxies form and how they evolve.

These mysterious objects, called *ultra-diffuse galaxies*, represent one of the most puzzling discoveries in modern astronomy. Unlike typical galaxies that shine brightly with billions of stars, these oddities are so faint and spread out that they are incredibly difficult to detect. They are like cosmic cotton candy, enormous in size but with almost no substance you can see.

The most famous example is Dragonfly 44, located about 100 million light years away in the Coma galaxy cluster. When astronomers first studied this ghostly galaxy, they made a shocking discovery: despite being roughly the same size as the Milky Way, Dragonfly 44 contains less than 1% of the stars our Galaxy has. If the Milky Way is like a bustling city blazing with lights, Dragonfly 44 is like a vast, nearly empty area of countryside with just a few scattered street lamps.

But here's where things get really strange. When scientists measured Dragonfly 44's gravitational effects on nearby objects, they found something impossible to ignore – this 'empty' galaxy actually has nearly the same total mass as the Milky Way. If it is not stars providing all that mass, what is?

The answer appears to be dark matter, the mysterious, invisible substance that makes up about 85% of all matter in the universe. While normal galaxies typically contain about five times more dark matter than regular matter, Dragonfly 44 is almost entirely dark matter. It seems to be held together by an invisible framework we can only detect through its gravitational influence.

This discovery raises profound questions about galaxy formation. How does a galaxy end up with so little star formation? Some theories suggest these ultra-diffuse galaxies might have formed in environments where gas was stripped away before it could collapse to form stars. Others propose they might be the remnants of larger galaxies that had their stars torn away by gravitational interactions with neighbouring galaxies.

Since the discovery of Dragonfly 44, astronomers have found hundreds of similar ultra-diffuse galaxies scattered throughout the universe. These ghostly galaxies remind us that the universe is far stranger and more diverse than we ever imagined. In a universe where we've grown accustomed to galaxies blazing with starlight, the discovery of dark, nearly starless galaxies opens entirely new chapters in our understanding of cosmic evolution and the hidden structures that govern our universe.

Dragonfly 44 captured by the Hubble Space Telescope. (NASA/ESA)

FACT 69

The Boomerang Nebula is the coldest place in the universe

Deep in space, about 5,000 light years from Earth, exists a place so cold it makes the empty vacuum of space feel toasty-warm by comparison. The Boomerang Nebula holds the record of being the coldest known place in the universe, with temperatures plummeting to just one degree above absolute zero, the theoretical point where all atomic motion ceases.

To put this in perspective, microwave background radiation – the echo from the Big Bang that permeates all of space – maintains a temperature of about -270° C. The Boomerang Nebula, however, reaches a bone chilling -272° C, making it even colder than the universe itself. This ghostly cloud of gas and dust has somehow managed to cool itself below the temperature of deep space, creating a cosmic refrigerator that defies our everyday understanding of how the universe works.

The nebula's extreme cold is not an accident. It is the result of a dying star's dramatic final act. At the heart of the Boomerang Nebula lies a red giant star that has begun ejecting its outer layers into space at tremendous speeds. These gas clouds are racing away from the central star at velocities exceeding 150 kilometres per second, roughly 336,000 miles per hour.

This high speed gas expansion is the key to the nebula's record-breaking low temperature. As the gas rushes outward, it undergoes rapid *adiabatic expansion* – a process similar to what happens when you release air from a pressurised spray can and feel it turn cold. The gas molecules spread out so quickly that they do not have time to absorb heat from surrounding space, causing the temperature to plummet dramatically.

The Boomerang Nebula earned its name from early observations that revealed a distinctive boomerang shape. However, later high resolution images from the Hubble Space Telescope showed the nebula actually resembles more of an hourglass or butterfly pattern. The characteristic shape results from the way gas and dust are being ejected from the dying star, likely influenced by the star's rotation and magnetic fields.

What makes this cosmic deep freeze even more remarkable is that it is only temporary on astronomical timescales. As the nebula continues to expand and the central star evolves, it will eventually warm up. The star is destined to become a white dwarf, and as it heats up, it will begin to ionise the surrounding gas, transforming the cold, dark nebula into a bright, colourful planetary nebula.

This frozen region reminds us that space contains environments far more extreme than anything we experience on our planet. While we bundle up against winter cold that might reach -40° C in some places, the Boomerang Nebula maintains temperatures that would instantly freeze any known form of matter, creating a realm where the normal rules of physics play out in their most extreme forms.

The Boomerang Nebula is a young planetary nebula and the coldest object found in the universe so far. (NASA/ESA)

FACT 70

Black holes can sing

In the vast silence of space, black holes are performing a cosmic concert that no human ear could ever hear. Black holes are known for devouring everything that comes too close, but they are also surprisingly musical, generating sound waves that resonate through the universe at frequencies so low they defy comprehension.

The discovery that black holes can produce sound seems to contradict everything we know about space. After all, space is a vacuum, and sound waves

270 hours of Chandra observations of the central regions of the Perseus galaxy cluster reveals evidence of the turmoil that has wracked the cluster for hundreds of millions of years from the supermassive black holes. (NASA/CXC/IoA/A.Fabian et al)

need matter to travel through. The famous movie quote, 'in space, no one can hear you scream' holds true for ordinary sound waves that vibrate through air. But black holes operate on a different scale entirely, creating acoustic waves that ripple through the hot gas and plasma that fills the space between galaxies.

The Perseus galaxy cluster, located about 240 million light years from Earth, provides the most famous example of a 'singing' black hole. At its centre lurks a supermassive black hole that has been humming the same note for an estimated 2.5 billion years, generating sound waves with a pitch over a million billion times deeper than humans can hear, specifically, a B-flat that sits 57 octaves below middle C on a piano.

To put this in perspective, the lowest note humans can typically hear has a frequency around 20 Hz, meaning the sound wave completes 20 cycles per second. The Perseus black hole's song operates at an unimaginably low frequency, taking hundreds of thousands of years to complete a single cycle. If you could somehow make it audible, it would sound like the deepest, most resonant bass note imaginable.

These sound waves are not just curiosities though – they play a crucial role in shaping the universe around them. As the black hole's gravitational pull draws in surrounding matter, it creates powerful jets of energy that push against the surrounding gas. This interaction generates pressure waves that ripple outward like sound waves, heating and stirring the intergalactic gas.

The discovery of these acoustic waves has changed our understanding of how black holes interact with their environments. Rather than being purely destructive forces, they act more like thermostats, regulating the temperature and density of gas in galaxy clusters. The sound waves help prevent the surrounding gas from cooling too quickly, which would otherwise trigger uncontrolled star formation.

The only way to detect these sound waves is by using space-based X-ray telescopes that can observe the ripples and variations in the hot gas surrounding black holes. These observations reveal characteristic patterns, concentric rings of brighter and dimmer gas that mark the passage of ancient sound waves.

The idea that black holes can sing challenges our perception of these mysterious objects. They are not just cosmic vacuum cleaners destroying everything in their path, but active participants in a grand orchestra, playing notes that help orchestrate the evolution of galaxies and the universe itself. In the silence of space, black holes remind us that even in the most extreme environments, the universe finds ways to make music.

FACT 71

The universe is getting darker

The universe is slowly dimming, and there's nothing we can do to stop it. Over billions of years, stars are burning out faster than new ones are being born, leading us inexorably toward a twilight that will eventually plunge the universe into permanent darkness.

This dimming is not happening overnight. It is a gradual process that unfolds over timescales almost impossible to comprehend. But the trend is clear: star formation in the universe peaked roughly 10 billion years ago and has been declining ever since. We are living through what astronomers call the universe's middle age, watching as the stellar nurseries that once blazed with newborn stars gradually go quiet.

The problem lies in the universe's expansion. As space itself stretches, the gas and dust needed to form new stars becomes increasingly diluted. Imagine trying to bake bread with flour that's constantly being spread thinner and thinner. Eventually you simply do not have enough concentrated ingredients to make anything substantial. The same thing is happening in the universe as the raw materials for star formation become scattered across ever larger volumes of space.

Meanwhile, existing stars continue their relentless march toward death. Our own Sun, a middle-aged star, has about 5 billion years of fuel remaining before it exhausts its hydrogen supply and begins its final transformation. Smaller red dwarf stars will burn much longer, some for trillions of years, but even these great lighthouses will eventually flicker out.

As we look toward the far future, astronomers predict we are heading toward what they call the 'degenerate era', a time when only the corpses of dead stars will remain. White dwarfs, neutron stars and black holes will be the universe's final inhabitants, slowly cooling in the darkness like embers from a once roaring fire.

This stellar graveyard will not arrive for an very, very long time though. Even the longest-lived red dwarf stars will continue burning for hundreds of trillions of years, a timespan so vast that if the entire history of the universe were compressed into a single year, we'd currently be living in the first few seconds of 1 January. But the universe's gradual darkening has already begun. Galaxies today produce only about 3% of the stars they did at the peak of star formation. We are witnessing the universe's slow fade from a brilliant youth filled with stellar fireworks to a dimmer, more subdued maturity.

This cosmological evolution reminds us how precious our current moment truly is. We exist during a brief golden age when the universe still sparkles with billions of active stars, when galaxies still spiral majestically across the sky, and when the night remains illuminated by countless points of light. Future civilisations, if any exist, will inhabit a much darker cosmos, making our current star-filled universe seem like a distant memory of better, brighter times.

The current era of star formation as captured here in the famous *Pillars of Creation* image from the Hubble Space Telescope will not continue for ever. (NASA)

FACT 72

There are natural telescopes in deep space

Hidden throughout the cosmos are invisible telescopes more powerful than any instrument humans have ever built. These natural magnifying glasses, created by the universe itself, use the strange physics of gravity to bend and focus light from the most distant corners of space, revealing galaxies that would otherwise remain forever hidden from view.

Abel 2218 is a great example of a gravitational lens. (Space Telescope Science Institute)

These natural telescopes are called *gravitational lenses*, and they work by exploiting Einstein's insight that massive objects actually warp the fabric of space and time. When light from a distant galaxy travels through this warped space around a massive object like a galaxy cluster, it gets bent and redirected, much as a glass lens bends light to create magnification.

The effect is so powerful that astronomers can see galaxies that formed when the universe was just a few hundred million years old. Without these natural telescopes, these ancient galaxies would be far too faint and distant for even our most sophisticated space telescopes to detect.

The process creates some of the most spectacular images in astronomy. When a massive galaxy cluster sits perfectly between Earth and a more distant galaxy, the foreground cluster acts like a magnifying glass, stretching and brightening the background galaxy's image. Sometimes this creates multiple images of the same distant galaxy, appearing as bright arcs or even complete rings of light around the lensing cluster, a phenomenon Einstein himself predicted but thought would be too rare to ever observe.

These 'Einstein rings' and gravitational arcs are not just beautiful, they are scientific goldmines. By studying how the light gets distorted, it is possible to map the distribution of dark matter in the lensing galaxy cluster, revealing just how the universe is held together. It is like using shadows to figure out the shape of an invisible object.

One of the most famous examples is the Hubble Space Telescope's observations of galaxy cluster Abell 2744, nicknamed 'Pandora's Cluster'. This gravitational lens has revealed dozens of background galaxies, some magnified by factors of 10 to 100 times their normal brightness. Without this natural magnification, these galaxies would remain completely invisible to our current technology.

The discovery of gravitational lensing has revolutionised our understanding of the early universe. It allow us to peer further back in time than would otherwise be possible, revealing how the first galaxies formed and evolved during the universe's cosmic dawn. Gravitational lensing has shown us that galaxy formation began much earlier and proceeded much faster than many scientists expected.

Perhaps most remarkably, gravitational lenses demonstrate that the universe provides its own tools for exploration. While we've spent billions of dollars building increasingly powerful telescopes, nature has been quietly operating cosmic observatories far more capable than anything we could construct. These invisible telescopes scattered throughout deep space continue to reveal new secrets about the origins of the universe.

FACT 73

We cannot see 27% of the universe

Imagine if 27% of everything around you was completely invisible, not just transparent, but utterly undetectable to your senses. This is not science fiction, it is the reality of our universe. More than a quarter of all matter in the cosmos is made of something called dark matter, a mysterious substance that refuses to interact with light in any way we can observe.

Dark matter detectors like the DEAP-3600 detector continue to probe the mysteries of dark matter. (Mark Ward)

This invisible matter does not emit, absorb or reflect light. It does not glow like stars, shine like planets, or even cast shadows. For all practical purposes, dark matter is the universe's ultimate hide-and-seek champion, remaining completely hidden from our most powerful telescopes.

The evidence for dark matter's existence comes not from seeing it directly, but from observing its gravitational effects on things we can see. When we study how galaxies rotate, we notice something puzzling – the outer edges of galaxies spin much faster than they should if only visible matter were providing the gravitational pull. It's like watching a figure skater spin faster when they should be slowing down – the physics does not add up unless there's additional mass we cannot see.

Galaxy clusters provide even more dramatic evidence. These massive collections of galaxies move and behave as if they contain far more matter than all their visible stars and gas combined. When scientists add up all the matter they can observe (every star, planet, gas cloud and speck of dust), they find it accounts for only about 5% of the universe's total mass and energy. The remaining 95% consists of dark matter and its even more mysterious cousin, dark energy.

The search for dark matter has become one of the greatest detective stories in science. Deep underground laboratories around the world house incredibly sensitive detectors, hoping to catch dark matter particles as they potentially pass through Earth. These experiments sit miles beneath the surface to shield them from cosmic rays and other interference, waiting patiently for the telltale sign of a dark matter collision.

Some scientists theorise that dark matter consists of exotic particles that interact only through gravity and the weak nuclear force. Others propose it might be composed of primordial black holes formed in the early universe. Despite decades of searching, dark matter's true nature remains one of the most perplexing mysteries in physics.

What makes this invisible universe even more mind-bending is that dark matter appears to form the structure upon which all visible matter hangs. Computer simulations suggest that without dark matter's gravitational influence, galaxies as we know them could not have formed. The stars, planets and everything we see exist only because of the invisible framework provided by this mysterious substance.

Living in a universe where more than a quarter of all matter is invisible humbles our human perspective. We pride ourselves on our technological achievements and scientific understanding, yet we remain blind to a fundamental component of reality. The universe keeps most of its secrets hidden in plain sight, reminding us how much we still have to discover about the cosmos we call home.

FACT 74

Galaxies are colliding in space

Across the universe, galaxies are locked in a crazy demolition derby, smashing into each other in spectacular collisions that reshape the very fabric of space. These galactic pile-ups, involving hundreds of billions of stars, play out over millions of years and fundamentally transform the landscape of our universe.

Despite the violent imagery, galaxy collisions are surprisingly gentle affairs. When two galaxies merge, individual stars rarely collide because the distances between them are so vast. If you could shrink our Sun to the size of a marble,

The Whirlpool Galaxy is a fabulous example of a pair of colliding, interacting galaxies. (NASA/ESA)

the nearest star would still be over 600 kilometres away. This means that during a galactic collision, stars mostly pass by each other like ships in the night, even as their parent galaxies undergo dramatic transformation.

Our own Milky Way Galaxy is currently heading toward just such an encounter. In around 4.5 billion years, we'll collide with our nearest major neighbour, the Andromeda Galaxy. This is not a distant theoretical event – astronomers can measure Andromeda approaching us at over 400,000 kilometres per hour. While this sounds terrifyingly fast, it will still take billions of years for the galaxies to actually meet and merge.

The collision process unfolds like a dance choreographed by gravity. As two galaxies approach, their gravitational fields begin to interact, distorting their once elegant spiral arms into chaotic streams of stars. Long tidal tails stretch out behind each galaxy which can extend for hundreds of thousands of light years, creating some of the most beautiful images in astronomy.

During the collision, the supermassive black holes at each galaxy's centre eventually spiral toward each other. When they finally merge, they release tremendous amounts of energy in the form of gravitational waves – ripples in space-time itself that we can now detect with instruments like LIGO (the Laser Interferometer Gravitational-Wave Observatory). This cosmic announcement signals the birth of a new, larger galaxy from the remains of two smaller ones.

These galactic mergers are not rare cosmic accidents. Many examples exist in the night sky and they are fundamental to how the universe evolves. Computer simulations suggest that virtually every large galaxy, including our own Milky Way, is the product of multiple ancient collisions. They also trigger spectacular bursts of star formation. As gas clouds from the merging galaxies compress and collide, they create ideal conditions for new stars to be born. Some of the most luminous and active galaxies in the universe are thought to be the products of recent galactic mergers.

When the Milky Way and Andromeda eventually collide, astronomers predict the result will be a giant elliptical galaxy. By then, our Sun will likely have exhausted its fuel, but if any observers remain, they will witness one of the universe's most magnificent spectacles – the creation of a new cosmic island from the merger of two ancient stellar cities.

FACT 75

There is a massive void in space

In the constellation Boötes lies one of the most unsettling discoveries in modern astronomy, a region of space so empty that it defies our understanding of how the universe should be structured. The Boötes Void, also known as the 'Great Nothing', is a spherical region of space approximately 330 million light years in diameter that contains almost no galaxies.

To grasp the enormity of this emptiness, the average region of space this size should contain about 2,000 galaxies. The Boötes Void contains fewer than 60. If our Milky Way Galaxy were located in the centre of this void, we would not have discovered the existence of other galaxies until the 1960s, fundamentally altering the course of astronomical history and our understanding of the universe.

The Bootes Void.

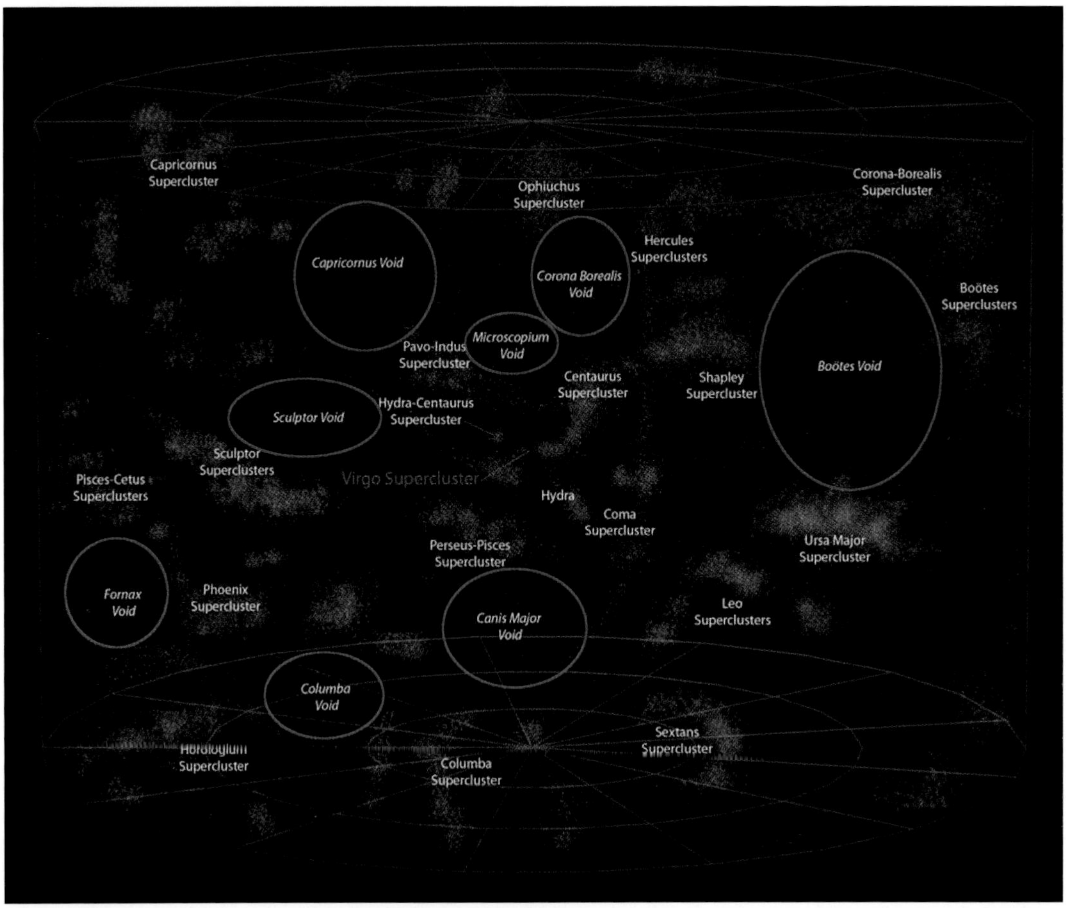

This strange cavity is so large that light, traveling at 300,000 kilometres per second, takes over 330 million years to cross from one side to the other. If you could somehow travel at the speed of light, you would spend more than three times the entire age of complex life on Earth just crossing this empty region.

The existence of the Boötes Void poses fundamental questions about the evolution of the universe. According to our current understanding, matter should be distributed relatively evenly on the largest scales, with galaxies forming along vast web-like structures of dark matter. Yet here sits a region where this cosmic web appears to have failed completely, leaving behind a bubble of near nothingness.

There are several theories to explain this anomaly. Some suggest it might be the result of multiple smaller voids merging over billions of years, gradually expanding and swallowing the galaxies that once occupied this space. Others propose it could be evidence of exotic physics from the universe's earliest moments, perhaps related to inflation or other poorly understood processes that shaped the universe after the Big Bang.

The few galaxies that do exist within the Boötes Void appear to be fleeing toward its edges, as if the emptiness itself is somehow repulsive. These galaxies are also unusually isolated, with vast distances separating them from any neighbours. For any hypothetical civilisations that might exist in these lonely galaxies, the night sky would be remarkably dark and empty, offering little evidence that a larger universe exists beyond their own galaxy.

The discovery of the Boötes Void has profound implications for our understanding of cosmic structure and evolution. It reminds us that the universe is far stranger and more varied than we might expect, containing not just dense clusters of galaxies but also vast regions of emptiness that challenge our theories about how matter organises itself on the largest of scales.

FACT 76

Astronauts can smell seared steak after a space walk

One of the most unexpected discoveries about space travel is not visible through telescopes or measurable by instruments, it is something you can only experience through your nose. Astronauts returning from spacewalks consistently report that their spacesuits carry a distinctive, pungent odour that many describe as smelling remarkably like seared steak, burnt metal or welding fumes.

This peculiar aroma is not something astronauts notice while floating in the vacuum of space itself. Since space is a vacuum, there's no air to carry odour molecules, making it impossible to smell anything during the actual spacewalk. The mysterious scent only becomes apparent when astronauts return to their spacecraft and remove their helmets in the pressurised environment of the airlock.

The smell is so consistent and distinctive that it has become part of astronaut folklore. NASA astronaut Don Pettit famously described it as 'a rather pleasant metallic sensation ... sweet smelling welding fumes'. Others have compared it to the smell of a hot metal workshop or, more appetisingly, to seared meat fresh off a grill. The aroma is strong enough that it can fill the entire airlock area, giving everyone aboard a whiff of space's unexpected signature scent.

Space is filled with cosmic radiation and charged particles from the solar wind, which constantly pelt any object venturing outside Earth's protective magnetic field. Scientists now believe this otherworldly odour results from high energy particles bombarding the spacecraft and spacesuits during spacewalks. When these energetic particles strike the materials, they cause complex chemical reactions at the molecular level.

The leading theory suggests that these high-energy collisions break apart molecules in the suit's materials, creating reactive fragments called *free radicals*. When astronauts bring their suits back into the pressurised environment, these freshly-created molecular fragments interact with oxygen and water vapour in the air, producing the characteristic odour that astronauts detect.

Different materials in the spacesuit may contribute to the complexity of the smell. The suit's outer layers contain various polymers, metals and protective coatings, each potentially contributing different notes to the overall aroma when altered by space radiation. It is like a cosmic chemistry experiment happening right on the astronauts' backs.

Interestingly, the smell does not seem to linger long. As the reactive molecules stabilise and the immediate chemical reactions subside, the distinctive scent gradually fades. This suggests that what astronauts are smelling is the immediate aftermath of ongoing space chemistry, not a permanent alteration of their equipment.

Astronaut Ed White during the first American spacewalk. (NASA)

The space-steak smell represents one of those wonderfully human aspects of space exploration, a sensory experience that connects our earthbound biology to the alien environment of space. While we can measure radiation levels and analyse cosmic particles with sophisticated instruments, it takes a human nose to reveal that space has its own distinctive perfume.

FACT 77

Neutron stars have mountains just millimetres tall

On the most extreme objects in the universe, even the tallest mountains are barely visible bumps. Neutron stars, the collapsed cores of massive stars, are so incredibly dense and have such powerful gravitational fields that their 'mountains' can only rise a few millimetres above the surface before the immense gravity crushes them back down.

To understand just how remarkable this is, imagine trying to build a mountain on an object where gravity is 200 billion times stronger than on Earth. On our planet, Mount Everest rises majestically to nearly 9 kilometres above sea level. But if you could somehow transport Everest to a neutron star, the crushing gravity would flatten it to a height barely thicker than a piece of paper in a fraction of a second.

Neutron stars are among the most extreme objects known to science. They pack more mass than our entire Sun into a sphere just 20 kilometres across. A teaspoon of neutron star material would weigh about 6 billion tons on Earth, equivalent to the mass of Mount Everest compressed into a coffee cup. The surface gravity is so intense that escape velocity, the speed needed to escape its pull of gravity, reaches about one third the speed of light.

Despite this crushing environment, neutron stars do have surface features. These millimetre-tall 'mountains' form through several processes. As neutron stars spin, some rotating hundreds of times per second, centrifugal forces can create tiny bulges. The star's intense magnetic field, trillions of times stronger than Earth's, can also warp the surface, creating small irregularities. Additionally, as neutron stars cool and their crusts solidify, thermal stresses can cause minute fractures and ridges to form.

These microscopic mountains may seem insignificant, but they do have effects on neutron star behaviour. As they spin, even millimetre-scale irregularities create gravitational waves, ripples in space-time itself that we can now detect with instruments like LIGO (the Laser Interferometer Gravitational-Wave Observatory), and these waves can cause the star to wobble slightly.

The discovery of these tiny surface features has changed our understanding of neutron star physics. We can now study how matter behaves under the most extreme conditions imaginable, where atomic nuclei are crushed together and normal physics breaks down. The crust of a neutron star is thought to be incredibly strong, possibly the strongest material in the universe, yet still yields to gravity's relentless pull.

Perhaps most remarkably, these millimetre-mountains represent the absolute limit of what solid matter can achieve in our universe. They exist at the boundary between what gravity will permit and what the fundamental forces holding matter together can resist. In a cosmos filled with towering peaks and vast mountain ranges, neutron stars remind us that sometimes the most extreme conditions produce the most humble landscapes.

The Crab Nebula including the central neutron star, it is the rightmost of the two bright stars near the centre of this image. (ESA/Hubble)

FACT 78

The universe has a beginning and will have an end

The universe is not eternal. Despite its seemingly infinite expanse and timeless appearance, our cosmos had a definitive beginning and will eventually reach a final conclusion. This revelation represents one of the most profound and arguably uncomfortable discoveries in human history, changing how we understand our place in the grand scheme of existence.

The beginning came 13.8 billion years ago in an event we call the Big Bang. This was not an explosion in space, but rather the explosive expansion *of* space itself from an infinitesimally small point containing all the matter and energy that has or would ever exist. In the first fraction of a second, the universe expanded faster than the speed of light, cooling from unimaginable temperatures as it grew. Within minutes, the first atomic nuclei formed, and

after hundreds of thousands of years, the first atoms appeared, allowing light to travel freely for the first time.

Evidence for all of this comes from multiple sources. We can observe the universe expanding today, with distant galaxies racing away from us in all directions. By running this expansion backward like a movie in reverse, we arrive at that singular moment of creation. We can also detect the afterglow of the Big Bang itself, the cosmic microwave background radiation that still permeates all of space, like the fading warmth from a fire that burned billions of years ago.

But if the universe had a beginning, logic demands it will also have an end. The nature of that ending depends on the fundamental properties of space, matter and the mysterious force called dark energy that's currently accelerating the universe's expansion.

The most likely scenario, based on current observations, is called the *heat death*. As space continues expanding, galaxies will drift apart until they disappear beyond each other's horizons. Star formation will eventually cease as the universe runs out of gas and dust. One by one, stars will burn out, leaving behind white dwarfs, neutron stars and black holes cooling in the darkness.

Even these stellar remnants will not last forever. Over incomprehensibly vast timescales, protons themselves will decay, causing all ordinary matter to disintegrate into elementary particles and radiation. The final act will see even black holes evaporate through Hawking radiation, leaving behind only a thin soup of particles and photons in an ever-expanding, ever cooling universe approaching absolute zero.

This endgame lies so far in the future that if the entire history of the universe were compressed into a single year, we would currently be living in the first second or so of 1 January. The heat death would not occur until countless aeons later, a timespan that dwarfs our ability to comprehend it.

Yet this finite nature of existence does not diminish the universe's majesty. Instead, it makes our current epoch, when stars shine, planets orbit and life contemplates its own existence, all the more precious. We exist during a brief, brilliant moment in history, witnesses to a universe that emerged from nothing and will eventually return to an even more profound emptiness.

This is not a printing error! The current theory for the end of the Universe is final state that is cold and dark as captured in this artist illustration. (Mark Thompson)

FACT 79

Most stars are red dwarfs

When we look up at the night sky, we see a dazzling array of bright stars twinkling against the darkness of space. But this view is deeply misleading. The vast majority of stars in our Galaxy (roughly 75%) are actually dim, cool red dwarf stars that are completely invisible to the naked eye. Our perception of the universe is skewed toward the rare, brilliant giants while missing the countless stars that truly dominate the stellar population.

Red dwarf stars are the quiet majority. These small, cool stars typically contain between 10 to 50% of our Sun's mass and burn at surface temperatures around 2,500 to 4,000°C, much cooler than our Sun's 5,500°C surface. This relatively low temperature gives them their characteristic reddish glow and makes them incredibly faint compared to stars like our Sun.

To understand just how dim these stars are, consider that the brightest red dwarf star visible from Earth, Proxima Centauri, requires a telescope to see despite being our very nearest neighbour at just 4.2 light years away. If our Sun were replaced by Proxima Centauri, Earth would be plunged into permanent twilight, receiving only about 0.0006% of the light we currently enjoy.

What red dwarfs lack in brightness, they make up for in longevity. These stellar equivalents to marathon runners are the ultimate survivors. While massive, brilliant stars burn through their fuel in just a few millions of years, red dwarfs conserve their hydrogen so efficiently that they can continue shining for hundreds of billions or even trillions of years. Some red dwarfs formed shortly after the Big Bang are still in their infancy, having barely begun their incredibly long lives.

This extraordinary lifespan has profound implications for life in the universe. Since red dwarfs will outlive virtually every other type of star, they represent the final frontier for life as we know it. Long after our Sun has died and most other stars have burned out, red dwarfs will continue providing energy and warmth to any planets orbiting them. In the far future, these humble stars may be the last refuges for life in an increasingly cold and darkening universe.

The dominance of red dwarfs also means that potentially habitable planets might be more common than we once thought. Many red dwarfs host rocky planets in their habitable zones, regions where liquid water could exist. While these worlds would experience perpetual sunset lighting and might be tidally locked to their stars, they could still potentially support life adapted to such conditions.

Proxima Centauri is the nearest red dwarf to our own solar system. (NASA/ESA)

FACT 80

Galaxies have habitable zones too

Just as planets must orbit at the right distance from their stars to support life, entire star systems must occupy the right neighbourhood within their galaxy to have a chance at developing complex life. This concept, known as the *galactic habitable zone*, suggests that our position in the Milky Way is not just lucky – it is essential for our very existence.

The idea emerged when astronomers realised that not all regions of a galaxy are equally hospitable to life. Galaxies are great big cities with distinct neighbourhoods, each presenting different challenges and opportunities for life to emerge and thrive. Some areas are like inner cities (crowded, energetic

Galaxies like the Sombrero Galaxy seem to have a habitable zone. (NASA/ESA)

and maybe dangerous) while others are more like the suburbs (quiet and stable, but perhaps lacking in essential resources).

Near the centre of a galaxy lies a chaotic realm dominated by a supermassive black hole surrounded by densely-packed stars. This region experiences frequent stellar explosions, intense radiation and gravitational disruptions that would make stable planetary systems nearly impossible. The radiation alone from supernovae and other energetic events would sterilise any nearby planets, stripping away their atmospheres and destroying any complex molecules necessary for life.

Venture too far toward the galactic edges and different problems emerge. These regions contain older, metal-poor stars that formed when the galaxy was young and had not yet been enriched with heavy elements. Planets forming around these ancient stars would lack the carbon, oxygen, silicon and iron necessary to build rocky worlds and complex chemistry. Without these heavier elements forged in previous generations of stars, life as we know it simply cannot develop.

Our solar system sits in what astronomers call the *Goldilocks zone* of the Milky Way. Here it is not too close to the dangerous galactic centre, not too far from the metal-rich star-forming regions, but just right. We orbit roughly 26,000 light years from the galactic centre, in a relatively quiet spiral arm where stellar density is moderate and the heavy element abundance is sufficient for rocky planet formation.

This galactic sweet spot provides several advantages for life. The stellar environment is stable enough that our solar system has remained relatively undisturbed for billions of years. We are far enough from the galactic centre to avoid frequent catastrophic events, yet close enough to benefit from the heavy elements produced by multiple generations of massive stars that lived and died before our Sun formed.

The concept of galactic habitable zones has implications for the search for extraterrestrial life. It suggests that even if we find planets in the habitable zones around their stars, their location within the galaxy matters enormously. A potentially life-bearing world orbiting a star near the galactic centre would face challenges that planets in our galactic neighbourhood never encounter. Our own existence may depend not just on Earth being the right distance from the Sun, but on the entire solar system being in the right neighbourhood of our Galaxy.

FACT 81

Space ripples like the ocean

Space itself can ripple like the surface of a pond disturbed by a falling stone. These waves, called *gravitational waves*, represent one of the most extraordinary discoveries in modern physics, proof that the very fabric of space-time can stretch and compress like a flexible material, carrying information about the universe's most violent events across billions of light years.

Einstein predicted these space ripples over a century ago as part of his theory of general relativity. He realised that massive objects do not just sit in space, they actually warp and curve space-time around them. When extremely massive objects accelerate or collide, they create disturbances that propagate outward at the speed of light, like waves spreading across the ocean.

The waves are almost impossibly subtle. When a gravitational wave passes through Earth, it stretches and compresses space by an amount smaller than 1/10,000th the width of a proton. To detect something this minuscule, scientists built the most sensitive measuring devices ever created: kilometre-long laser interferometers that can detect changes in distance smaller than anything previously imagined.

The Laser Interferometer Gravitational-Wave Observatory (LIGO) consists of two massive L-shaped detectors, one in Louisiana and one in Washington state. Each arm of these detectors is 4 kilometres long, and laser beams constantly measure the distance between mirrors at each end. When a gravitational wave passes through, it causes one arm to lengthen slightly while the other shortens, creating a measurable difference in the laser travel times.

On 14 September 2015, LIGO detected its first gravitational wave signal lasting just two-tenths of a second. This brief signal carried the story of two black holes, each about 30 times more massive than our Sun, spiralling around each other and finally merging in a catastrophic collision that occurred 1.3 billion years ago. The energy released in those final moments was greater than the light output of all the stars in the observable universe combined.

Since that first detection, scientists have observed dozens of these ripples, each telling a unique story. Most come from black hole mergers, but in 2017 LIGO detected waves from colliding neutron stars, an event that also produced a brilliant flash of light visible to conventional telescopes. This marked the beginning of 'multi-messenger astronomy', where scientists can study cosmic events using both light and gravitational waves.

These space ripples have opened an entirely new window on the universe. Just as the invention of the telescope revolutionised astronomy by revealing previously invisible light, gravitational wave detectors allow us to 'hear' the universe in ways never before possible. The cosmos is no longer silent. It is filled with a symphony of space-time vibrations from some of the most extreme events imaginable.

Gravity waves in space are like waves on the ocean, captured here in the Timor Sea, Western Australia. (Jeff Schmaltz)

FACT 82

The largest black hole is as massive as 17 billion suns

In the heart of a distant galaxy lurks a monster so massive it defies comprehension. The supermassive black hole at the centre of galaxy NGC4889, a name that rather rolls off the tongue, located roughly 300 million light years from Earth, contains the mass of 17 billion suns compressed into a region smaller than our solar system! This cosmic leviathan represents the upper limit of what black holes can achieve, pushing the boundaries of physics and challenging our understanding of how such giants could even form.

To grasp the enormity of this monster, imagine trying to fit 17 billion copies of our Sun into a space roughly the size of Pluto's orbit. The density becomes so extreme that the very concept of matter as we know it breaks down. This black hole's event horizon, the point of no return where not even light can escape, stretches approximately 130 billion kilometres across, making it large enough to swallow our entire solar system with room for several more.

The gravitational influence of such a massive object is equally staggering. From Earth's distance, this black hole's gravity would appear about 3,000 times stronger than our Sun's. If you could somehow replace our Sun with this black hole while maintaining the same distance, Earth would be pulled into the black hole in mere minutes, accelerating to incredible speeds as space-time itself warped around the approaching monster.

Galaxy NGC4889 which is host to a supermassive black hole. (NASA/Hubble)

What makes these ultra massive black holes particularly mysterious is how they could have grown so large. Black holes typically form when massive stars collapse, but even the most massive stars produce black holes only dozens of times heavier than the Sun. For a black hole to reach 17 billion solar masses, it would need to devour enormous amounts of matter continuously for billions of years, essentially eating the equivalent of several entire galaxies.

Current theories suggest these giants may have formed in several ways. Some might have started as smaller 'seed' black holes in the early universe that grew by constantly feeding on surrounding gas and stars. Others could have formed from the direct collapse of massive gas clouds in the primordial universe, giving them a head start on their journey to become supermassive.

The existence of such massive black holes also plays a crucial role in galaxy formation and evolution. These giant black holes regulate star formation in their host galaxies through powerful jets of energy that heat the surrounding gas, preventing it from cooling and forming new stars. Perhaps most remarkably though, black holes of this magnitude may represent a natural upper limit to how massive these objects can become. As they grow larger, they become increasingly efficient at heating and dispersing the very material they need to continue growing, potentially starving themselves once they reach a certain size.

FACT 83

Nothing can travel faster than the speed of light, except the universe

Einstein's speed limit is absolute, nothing with mass can travel faster than light (299,792,458 metres per second). This fundamental law governs everything from subatomic particles to spaceships, creating an insurmountable barrier that

Edwin Hubble discovered the Universe was expanding and could be detected by studying the light of distant galaxies. (Johan Hagemeyer)

prevents us from achieving instantaneous travel across the universe. Yet the universe itself laughs at this restriction, expanding so rapidly during its early stages that distant galaxies are racing away from us faster than light itself can travel.

This contradiction, fantastical as it sounds, is not science fiction, it is the mind-bending reality of our expanding universe. When we say the universe is expanding, we do not mean galaxies are flying through space at incredible speeds like rockets. Instead, space itself is stretching, carrying galaxies along for the ride like raisins in rising bread dough. The raisins are not moving through the dough; the dough itself is expanding and taking them with it.

Einstein's speed limit applies to objects moving through space, but it says nothing about space itself expanding. This crucial distinction allows the universe to perform what seems like magic: galaxies billions of light years away are receding from us at velocities that exceed the speed of light, yet they are not actually violating relativity because they are not moving through space, space is expanding between us and them.

The observable consequences of this expansion are profound. There are galaxies so distant that their light will never reach us, no matter how long we wait, so we can never see them! The space between us and these galaxies is expanding faster than light can cross it, creating a horizon beyond which we can never see. It is like trying to swim upstream in a river flowing faster than you can swim – you'll never make progress toward your destination.

This expansion began during the universe's inflation period, a brief time immediately after the Big Bang when the universe expanded faster than light for a tiny fraction of a second. During this growth spurt, regions of space that were once close neighbours were suddenly separated by vast distances, all without violating Einstein's speed limit, because no matter was actually moving through space at *superluminal* or faster-than-light speeds.

Today, we can observe this ongoing expansion through the red shift of distant galaxies. As space stretches between us and far-off galaxies, it stretches the wavelengths of light travelling through it, making the light appear redder when it reaches us. The more distant the galaxy, the faster space is expanding between us, and the more dramatically redshifted its light becomes.

Perhaps most remarkably, this expansion is accelerating. Dark energy, a mysterious force comprising about 68% of the universe, is pushing space to expand ever faster. This means that in the far future, galaxies will disappear beyond our cosmic horizon one by one, leaving future astronomers in an increasingly empty and isolated observable universe.

FACT 84

Our solar system is surrounded by a bubble

Our solar system exists inside a giant bubble, floating through space like a protective sphere that shields us from the harsh radiation of interstellar space. This invisible barrier, called the *heliosphere*, is created by our Sun and extends far beyond the orbit of Pluto, forming a magnetic cocoon that has quietly protected life on Earth for billions of years.

This illustration shows the position of NASA's *Voyager 1* and *Voyager 2* probes, outside the heliosphere – a protective bubble created by the Sun that extends well past the orbit of Pluto. (NASA/JPL Caltech)

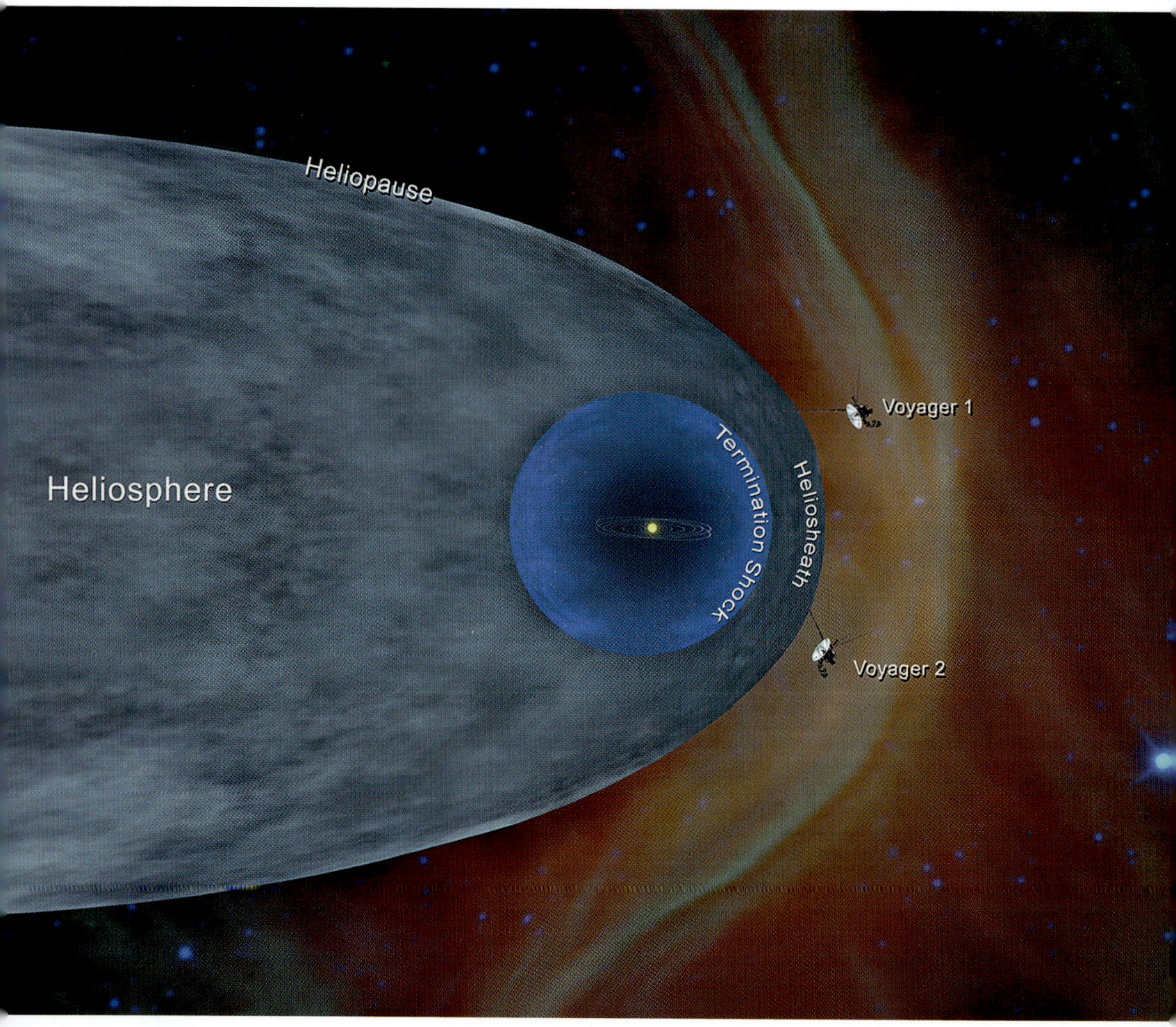

The heliosphere forms when the solar wind, a constant stream of charged particles flowing from the Sun at over a million miles per hour, collides with the *interstellar medium*, the thin soup of gas and dust that fills the space between stars. The Sun's particle stream pushes against the surrounding interstellar material, carving out a bubble-shaped cavity approximately 100 times wider than the distance from the Earth to the Sun.

The protective heliosphere is not perfectly round. As our solar system moves through the Galaxy at about 80,000 kilometres per hour, the heliosphere gets compressed on the leading edge and stretched into a long tail on the trailing side, making it look more like a comet than a sphere. The boundary where the solar wind pressure balances the pressure of interstellar space is called the *heliopause*, marking the edge of the Sun's direct influence.

The importance of this stellar bubble cannot be overstated. Beyond the heliosphere lies a much more dangerous environment filled with high-energy cosmic rays, charged particles accelerated to incredible speeds by distant supernovae and other violent events. These cosmic rays can damage DNA, disrupt electronics and strip away planetary atmospheres. Our heliosphere acts like a shield, deflecting about 70% of these harmful particles away from the inner solar system.

Without this protection, life on Earth would face a constant barrage of dangerous radiation. Thankfully, the heliosphere's magnetic field deflects most incoming rays, allowing only the lower-energy particles to penetrate our neighbourhood. This shielding may have been crucial for the development and survival of life on our planet, providing a relatively stable radiation environment for billions of years.

The *Voyager* spacecraft have provided our best glimpse of this boundary. *Voyager 1* crossed the heliopause in 2012, becoming the first human-made object to enter interstellar space. It immediately detected a dramatic increase in cosmic ray intensity and changes in the magnetic field, confirming that it had indeed left our solar system's protective bubble. *Voyager 2* followed suit in 2018, providing additional data about this invisible frontier.

Interestingly, other stars create their own heliospheres, suggesting that stellar bubbles are common throughout the Galaxy. Some stars produce much larger bubbles than our Sun, while others create smaller or differently-shaped protective zones depending on their size, age and the local interstellar environment they are moving through.

FACT 85

Water ice has been found on an exoplanet

For the first time in history, astronomers have detected water ice clouds in the atmosphere of an exoplanet, marking a milestone in our search for potentially habitable worlds beyond our solar system. This discovery, made using the James Webb Space Telescope (JWST), reveals that planets orbiting distant stars can harbour the same fundamental ingredients for life that we find on Earth.

The discovery was made on WASP-96b, a gas giant located approximately 1,150 light years from Earth in the constellation Phoenix. While this particular planet is far too hot and massive to support life as we know it, with temperatures exceeding 1,000° C, the detection of water ice in its atmosphere represents a technological breakthrough that opens the door to studying potentially habitable worlds.

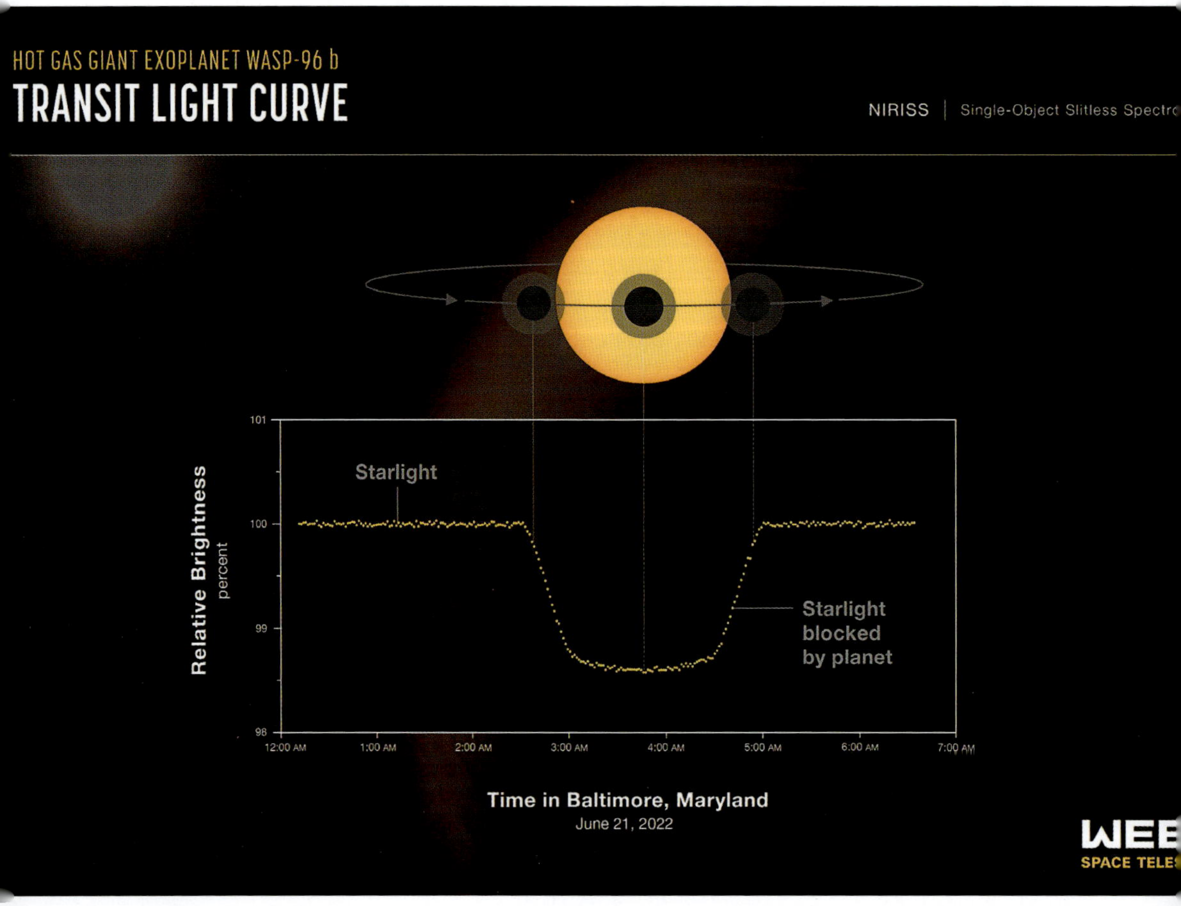

WASP-96b is what is known as a 'hot Jupiter', a type of gas giant that orbits extremely close to its host star. Despite the scorching temperatures at lower altitudes, the planet's upper atmosphere contains regions cold enough for water vapour to condense into ice crystals, forming clouds similar to those we see in Earth's atmosphere, albeit under dramatically different conditions.

The detection was made possible by analysing starlight as it passed through the planet's atmosphere during a *transit*, when the planet crosses in front of its star as viewed from Earth. As light filters through the atmospheric layers, different molecules absorb specific wavelengths, creating a unique spectroscopic fingerprint. Water ice has a distinctive signature that the JWST's incredibly sensitive instruments can detect and analyse.

This achievement represents more than just finding frozen water on a distant world. It demonstrates that we now possess the technology to analyse the atmospheric composition of exoplanets in amazing detail. The same techniques used to find water ice on WASP-96b can be applied to smaller, cooler planets that might exist in their star's habitable zone, the region where liquid water could exist on a planet's surface.

The implications for astrobiology are profound. Water is considered essential for life (especially life like ours,) and its presence in any form on exoplanets suggests that the building blocks for life might be more common throughout the Galaxy than previously thought. While ice clouds alone do not indicate the presence of life, they represent a crucial step toward identifying worlds where life could potentially exist.

This discovery also validates decades of theoretical work about exoplanet atmospheres. Scientists had predicted that water-ice clouds should exist on certain types of exoplanets, but confirming this prediction required instruments sensitive enough to detect the subtle spectroscopic signatures across vast interstellar distances.

A light curve from the James Webb Space Telescope that shows the change in brightness of light from the WASP-96 star system over time as the planet transits the star.
(NASA, ESA, CSA and STScI)

FACT 86

Planets are forming at the centre of our Galaxy

Life finds a way and, apparently, so do planets. In one of the most surprising astronomical discoveries in recent years, astronomers have found that planets are actively forming in the most hostile and chaotic region of our Galaxy, right near the supermassive black hole at the Milky Way's heart. This discovery overturns decades of scientific assumptions about where worlds can be born.

For years, the galactic centre was considered the last place you'd expect to find planet formation. This region is a maelstrom of violence: intense

Two of the Atacama Large Millimeter/submillimeter Array (ALMA) 12-metre antennas gaze at the sky at the observatory's Array Operations Site (AOS), high on the Chajnantor plateau at an altitude of 5000 metres in the Chilean Andes. (Iztok Bončina/ESO)

radiation bathes everything in lethal doses, massive stars explode regularly as supernovae and powerful gravitational forces from our Galaxy's central black hole tear apart anything that ventures too close. The environment seemed far too turbulent and destructive for the delicate process of planet formation to occur.

This assumption was wrong. Observations from the Atacama Large Millimeter Array (ALMA) have revealed something extraordinary – protoplanetary discs where planets form are not just surviving but actually thriving in the turbulent Central Molecular Zone (CMZ) near our Galaxy's heart. These planet-forming discs are flourishing in conditions that should theoretically make their existence impossible.

Protoplanetary discs are swirling clouds of gas and dust that surround young stars. Over millions of years, particles within these discs gradually stick together, growing from microscopic grains to pebbles, then to boulders and eventually to full-sized planets. The process requires relatively stable conditions for dust particles to slowly accumulate without being blown apart by stellar winds or radiation.

The galactic centre presents exactly the opposite conditions. Stellar winds are fierce, radiation levels are extreme and the gravitational environment is chaotic. Yet somehow, these crazy construction sites are managing to build worlds despite the odds stacked against them. It is like discovering that delicate flowers are growing in the middle of a hurricane!

The discovery challenges our fundamental understanding of planetary formation. If planets can form in the Galaxy's most extreme environment, it suggests that world-building is far more resilient and adaptable than scientists previously believed. The universe appears to be more creative and persistent in its planet-making endeavours than we ever imagined.

These findings have profound implications for the prevalence of planets throughout the universe. If planetary systems can emerge even in the most hostile galactic neighbourhoods, then planets might be far more common than current estimates suggest. Regions of the galaxy we previously wrote off as uninhabitable wastelands might actually be teeming with worlds.

The discovery also raises intriguing questions about the types of planets that might form in such extreme conditions. Would worlds born in the galactic centre's crucible be fundamentally different from planets like Earth? Could they possess unique characteristics that help them survive in high radiation environments? Further research is needed but the concept is tantalising.

FACT 87

Black holes are firing gas out into space

Black holes have a reputation as vacuum cleaners in space that devour everything in their path, but new research reveals they are also the universe's most powerful cannons. It has been discovered that these gravitational monsters do not just consume matter, they also fire it back with tremendous force, launching high-speed gas projectiles into space at speeds that would make *Star Wars* blasters jealous.

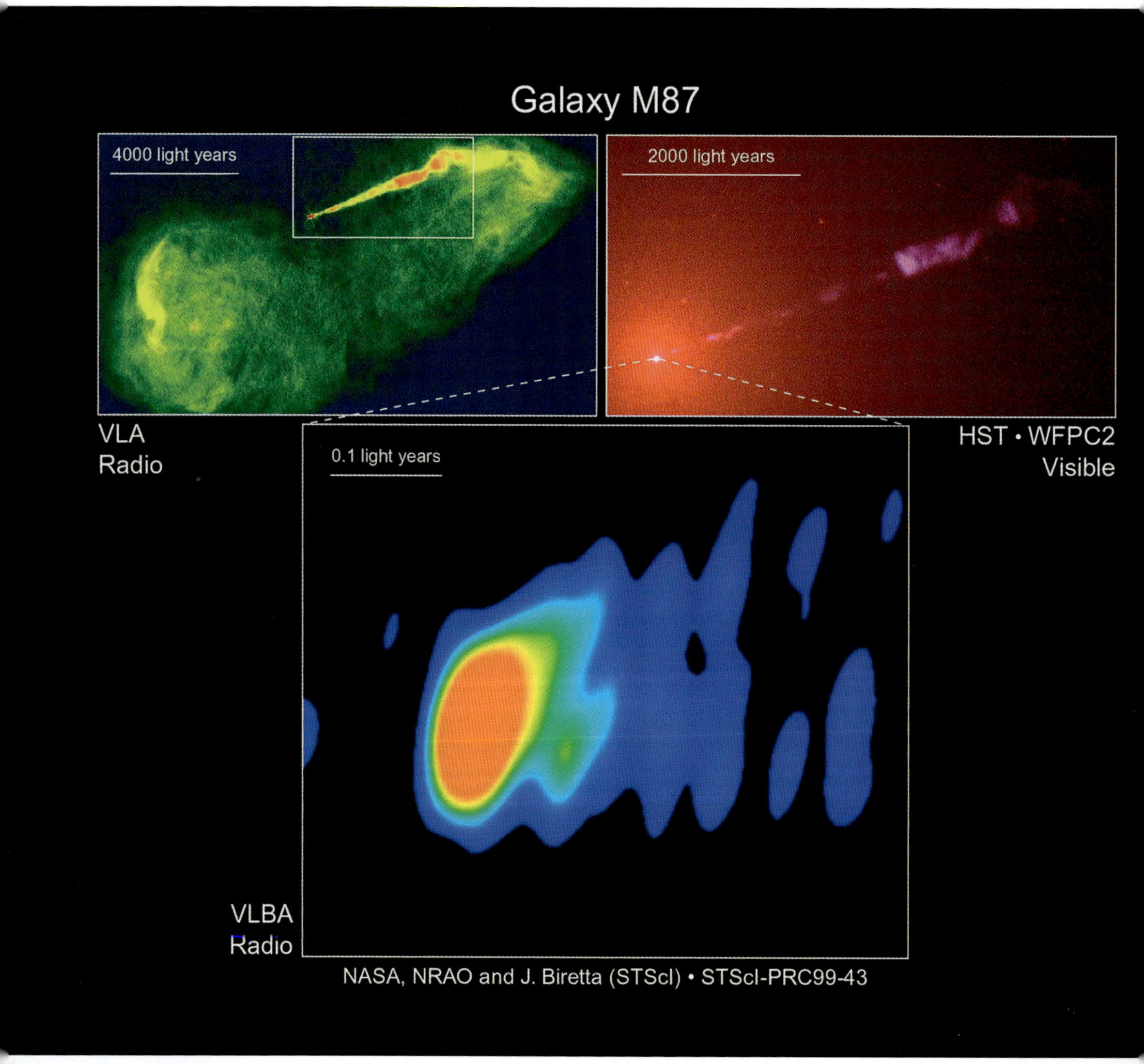

While nothing can escape once it crosses the event horizon, the point of no return around a black hole, the region just outside this boundary is surprisingly violent and dynamic. As matter spirals toward the black hole, it forms a superheated disc that glows brilliantly and generates ferocious winds. These winds blast outward at significant fractions of the speed of light, creating some of the most energetic phenomena in the universe.

It is been assumed for decades that these outflows were smooth and continuous, like water flowing steadily from a garden hose. New research has changed this assumption, revealing that black-hole winds are actually violent, fragmented bursts that resemble rapid-fire streams of gas bullets.

Recent observations have captured this phenomenon in unprecedented detail, detecting five distinct gas components traveling at 20–30% the speed of light, roughly 60,000 to 90,000 kilometres per second. To put this in perspective, these gas bullets could travel from Earth to the Moon in just over a minute, while it took Apollo astronauts about four days to reach the Moon. The gas erupts like geysers from the black hole's vicinity, creating a spectacular light show visible across vast distances.

These high-speed outflows are not just impressive displays of power though, they play a crucial role in shaping the universe around them. The energy carried by these gas streams is so enormous that it can halt star formation in entire galaxies, regulate the growth of the black hole itself and even strip gas from neighbouring galaxies millions of light years away.

The discovery helps solve how supermassive black holes manage to regulate their own growth and influence their host galaxies so dramatically. The answer lies in these violent outbursts that act like pressure valves, preventing the black hole from growing too large too quickly, while simultaneously affecting the evolution of entire galactic neighbourhoods.

Scientists detected these gas bullets using sophisticated X-ray telescopes that can peer through the dusty environments around black holes. The gas becomes so hot during its violent ejection, reaching temperatures of millions of degrees, that it glows brightly in X-ray light, creating distinctive signatures that astronomers can analyse to understand the physics of these extreme events.

Jets seen coming from the black hole at the centre of the galaxy M87 (NASA)

FACT 88

Stars can orbit inside other stars

In one of the most bizarre discoveries in astronomy, scientists have found a star system where one star literally orbits inside the atmosphere of another. This strange oddity, named PSR J1928+1815, challenges our understanding of how stellar systems can exist, revealing that the universe has far stranger arrangements than we ever imagined possible.

The extraordinary system features a pulsar – a rapidly spinning neutron star that acts like a lighthouse beaming radio signals across space with clockwork precision. This stellar remnant is locked in an incredibly tight dance with a companion star, completing a full orbit in just 3.6 hours. To put this in perspective, while our Moon takes 28 days to orbit Earth, these two stars whip around each other faster than most people's day at work.

The orbit is so close that the pulsar actually plunges through the outer atmosphere of its companion star during each cycle. From our vantage point on Earth, we can observe this as an eclipse as the pulsar's normally regular radio signals vanish for roughly one-sixth of each orbit, about 36 minutes, when they are blocked by the companion star's interference. It is like watching a torch beam disappear as it passes behind a tree, except the 'tree' is another star's atmosphere.

This bizarre arrangement did not happen overnight. Astronomers believe they are witnessing the aftermath of a stellar catastrophe that unfolded over thousands of years. Originally, this was a normal binary system with two massive stars orbiting each other at a respectable distance. The more massive star, burning through its fuel faster, reached the end of its life first and exploded as a supernova, one of the most violent events in the universe.

The explosion was so powerful that it should have torn the binary system apart, sending both stars flying in opposite directions. Instead, something remarkable happened. The core of the exploded star collapsed into an ultra-dense neutron star, while the blast somehow brought the two stellar remnants closer together rather than pushing them apart.

The newly-formed neutron star found itself spiralling inward, eventually passing within the bloated atmosphere of its companion. For approximately 1,000 years, the neutron star carved a path through the outer layers of its partner, gradually blowing away the envelope of material that surrounded both stars. This orbital arrangement slowly stripped away the companion's outer atmosphere, creating the tight, intimate orbit we observe today.

The discovery reveals that stellar systems can survive and evolve in ways that seem almost impossible. The fact that a neutron star can orbit within another star's atmosphere without destroying either object demonstrates the incredible resilience of stars under extreme conditions.

Pulsar B1509-58 visible in X-ray light from Chandra space telescope. Pulsars create powerful beams that act like celestial lighthouses. (NASA/CXC/SAO)

FACT 89

The Hercules-Corona Borealis Great Wall is the largest structure in the universe

Stretching across an almost incomprehensible 10 billion light years of space, the Hercules-Corona Borealis Great Wall stands as the largest known structure in the observable universe. This megastructure is so vast that light, travelling at 300,000 kilometres per second, takes 10 billion years to cross from one end to the other. To put this in perspective, this structure is roughly one tenth the diameter of the entire observable universe.

The Great Wall is not a solid barrier though, like its earthly namesake, but rather a massive collection of galaxy clusters, superclusters and filaments of dark matter arranged in a web-like pattern.

Discovered in 2013 through careful analysis of gamma-ray bursts (GRBs), the most energetic explosions in the universe since the Big Bang, this structure initially seemed too large to exist according to our understanding of the evolution of the universe. The discovery team noticed that these incredibly bright flashes of light were not randomly distributed across the sky but instead showed a distinct clustering pattern in the constellations Hercules and Corona Borealis.

The existence of such a massive structure poses fundamental challenges to cosmology. According to the cosmological principle, matter should be distributed relatively uniformly on the largest scales. The Great Wall violates this assumption, suggesting that the universe has organised itself into structures far larger than current theories predict should be possible.

To grasp the true enormity of this giant, consider that our entire Local Group of galaxies, including the Milky Way, Andromeda and dozens of smaller galaxies, would appear as little more than a speck of dust compared to the Hercules-Corona Borealis Great Wall. Even our local supercluster, containing thousands of galaxies, represents just a tiny fraction of this incredible megastructure.

The Wall's discovery forces astronomers to reconsider how the universe evolved from the relatively smooth conditions shortly after the Big Bang to the complex web of structures we observe today. If such massive formations can exist, it suggests that gravity has been more efficient at organising matter on cosmic scales than previously thought, or that unknown forces have shaped the universe's architecture.

The structure also provides insights into the mysterious dark matter that comprises about 27% of the universe. The Great Wall likely formed along vast filaments of dark matter that act as scaffolding, guiding ordinary matter into these enormous configurations over billions of years.

Perhaps most remarkably, we are observing this structure as it existed billions of years ago, when the universe was much younger. The light we see today began its journey when the cosmos was still forming its largest structures, offering us a window into the universe's architectural history.

A massive galaxy cluster named MACS J0717.5+3745, which may be similar to the Hercules–Corona Borealis Great Wall. (NASA/ESA)

FACT 90

There's a quasar that outshines 600 trillion suns

In the depths of space lurks the brightest object ever discovered by human astronomers, a cosmic lighthouse so brilliant it makes our Sun look like a flickering candle. Quasar J0529-4351, discovered in 2024, blazes with the luminosity of 600 trillion Suns, representing the most luminous beacon in the known universe.

To grasp the sheer magnitude of this celestial powerhouse, imagine replacing our Sun with this quasar. The light would be so intense it would instantly vaporise Earth and every planet in our solar system. The energy output is so enormous that it could power human civilisation for trillions upon trillions of years, using our current global energy consumption as a baseline.

This image shows the region of the sky in which the record-breaking quasar J0529-4351 is situated. (ESO/Digitized Sky Survey 2)

This extraordinary brightness comes from one of the universe's most extreme phenomena, a supermassive black hole in the process of devouring everything around it. At the heart of J0529-4351 lies a gravitational monster so massive that it warps space-time itself, creating a whirlpool that pulls in vast amounts of matter from its surroundings.

As gas, dust and even entire stars spiral toward their doom, they form an accretion disc, a superheated maelstrom of matter orbiting the black hole at incredible speeds. The friction generated heats the material to temperatures exceeding millions of degrees, causing it to glow with an intensity that outshines entire galaxies.

What makes J0529-4351 particularly remarkable is its sheer efficiency. The supermassive black hole at its centre is consuming matter at a rate that challenges our understanding of physics. Every second, this monster devours material equivalent to hundreds of Earth masses, converting a portion of that matter directly into pure energy according to Einstein's famous equation $E = mc^2$.

The discovery of such an intensely luminous object provides astronomers with a unique laboratory for studying black hole physics under the most extreme conditions imaginable. Quasars like J0529-4351 played a crucial role in the evolution of the universe. Their intense radiation helped clear away the fog of neutral hydrogen that filled space after the Big Bang, a process called *reionisation* that made the universe transparent to light and enabled the formation of the cosmic structures we see today.

The existence of such an incredibly bright object also raises questions about the limits of luminosity. If a single object can outshine 600 trillion stars, what does this tell us about the maximum brightness possible in our universe? Are there even brighter objects waiting to be discovered in the cosmic depths?

FACT 91

Cosmic rays can mutate human beings

Every second of every day, invisible bullets from space are silently penetrating your body and potentially altering your DNA. Cosmic rays, high energy particles born in the violent explosions of distant stars, are constantly bombarding Earth and everything on it, including human beings. These subatomic projectiles pack

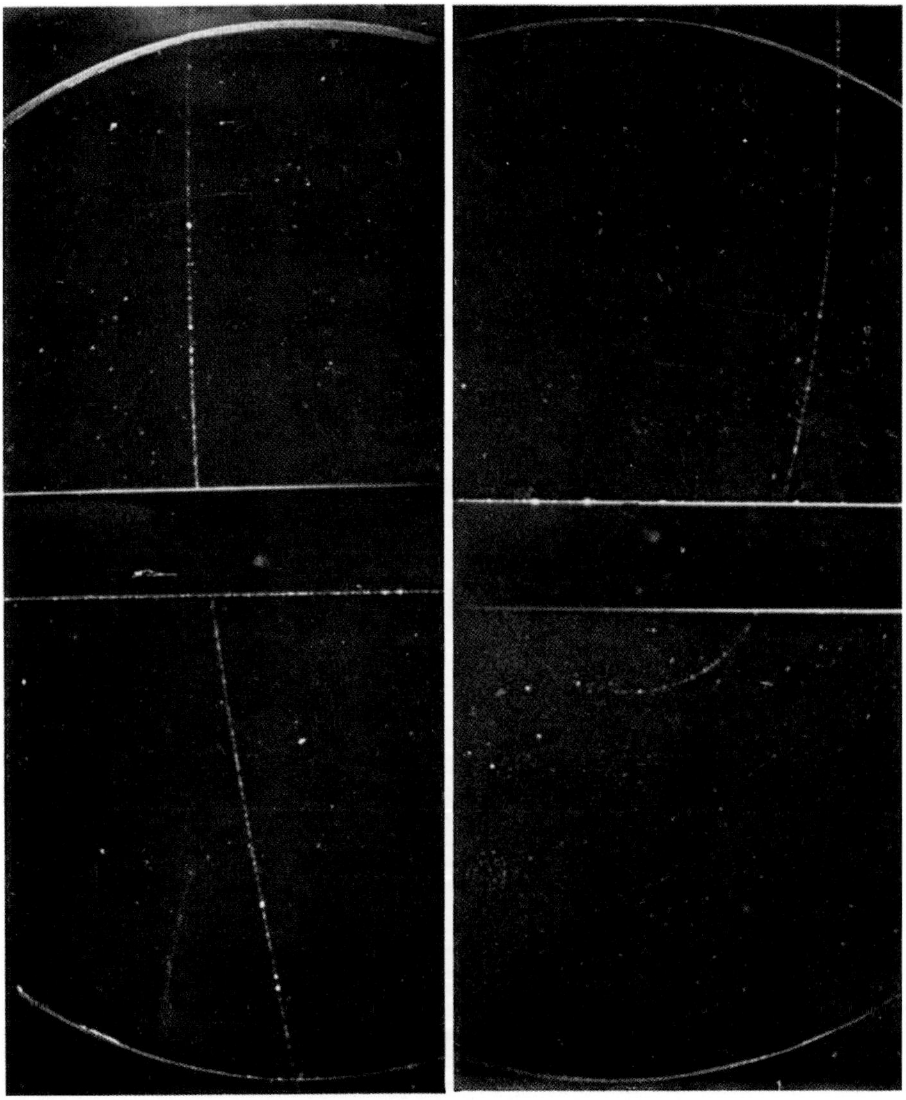

Cosmic rays can be seen in this image of a cloud chamber as they interact. (Jino john1996)

enough energy to punch through solid rock, and when they strike the delicate molecular machinery inside our cells, they can cause mutations that may affect our health, evolution and even our future as a species.

Cosmic rays originate from some of the most extreme events in the universe: supernovae, colliding neutron stars and the vicinity of supermassive black holes. These catastrophic events accelerate particles to nearly the speed of light, sending them careening across the Galaxy like microscopic bullets fired from stellar cannons. When these particles reach Earth, they create cascades of secondary particles that rain down on everything below.

The human body offers little protection against these invaders from space. A typical cosmic ray carries millions of times more energy than the particles produced by medical X-ray machines. While most pass harmlessly through our bodies, some inevitably collide with the atoms in our DNA, potentially breaking chemical bonds or causing structural damage to our genetic code.

The effects can be surprisingly significant. It is possible to estimate that cosmic rays cause approximately one DNA mutation per cell per year in the average human being. Over a lifetime, this translates to millions of genetic alterations throughout our bodies. Most of these mutations are harmless or quickly repaired by our cellular defence mechanisms, but some can trigger cancer, accelerate ageing or cause other health problems.

Airline pilots and flight attendants face particularly high exposure to cosmic radiation due to their frequent high altitude flights, where the atmosphere provides less protection. A pilot flying international routes receives roughly the same annual radiation dose as a nuclear power plant worker. Some studies suggest this occupational exposure may increase their risk of certain cancers and other radiation-related health issues.

Space travellers face even greater risks. Beyond Earth's protective magnetic field and atmosphere, astronauts experience cosmic ray bombardment hundreds of times more intense than what we experience on the ground. During long missions to Mars or other distant destinations, this radiation exposure could cause significant DNA damage, potentially increasing cancer risks and affecting fertility.

Interestingly, cosmic rays may have played a crucial role in human evolution. Over millions of years, cosmic ray-induced mutations may have contributed to the genetic diversity that allowed our species to adapt and evolve. Some scientists even theorise that periods of increased cosmic ray exposure, possibly triggered by nearby supernovae, could have accelerated evolutionary changes in early human ancestors.

FACT 92

It is unusual, but some stars form in complete isolation

In the vast emptiness between galaxies, far from any stellar nursery, some stars are born completely alone. Forming in the most unlikely places imaginable they challenge everything astronomers thought they knew about how stars come into existence. While most stars are born in crowded stellar nurseries surrounded by hundreds or thousands of siblings, these isolated stars manage to ignite in the vast emptiness of space with no neighbours for millions of light years.

The discovery of these solitary stars has puzzled astronomers for decades. According to conventional theory, star formation requires dense clouds of gas and dust that are typically found in the spiral arms of galaxies or in active star-forming regions. These stellar nurseries provide the perfect conditions for gravity to compress material until it becomes hot and dense enough to trigger nuclear fusion, the process that makes stars shine.

Yet somehow, stars are forming in the vast voids between galaxies, regions so empty that if our solar system were located there, the nearest galaxy would appear as little more than a faint smudge in our most powerful telescopes. These intergalactic spaces contain roughly one atom per cubic metre, making them emptier than the best vacuum chambers we can create on Earth.

So how do these lone stars come to exist? Scientists have proposed several intriguing explanations. One possibility is that these stars are not actually forming where we see them. Instead, they might be stellar refugees, stars that were born in normal galaxies but were later ejected through gravitational interactions with massive black holes or during galactic collisions. These stellar exiles could travel for billions of years through empty space, eventually finding themselves in the space between galaxies.

Another theory suggests that these isolated stars might form from primordial gas clouds left over from the early universe. During these dark ages, before the first galaxies formed, the universe was filled with vast reservoirs of hydrogen and helium. Some of these ancient gas clouds might have remained relatively undisturbed for billions of years, slowly cooling and condensing until they finally reached the critical density needed for star formation.

Perhaps most remarkably, some of these isolated stars might be forming from material stripped away during galactic encounters. When galaxies pass close to each other, their gravitational fields can tear away long streamers of gas and dust, flinging this material into intergalactic space. Over time, these orphaned clouds might accumulate enough mass to collapse and form new stars in complete isolation.

The existence of isolated stars has implications for our understanding of stellar demographics. If stars can form in the emptiest regions of space, it suggests that the universe might contain far more stars than we currently estimate. These hidden populations of isolated stars could be lurking in the darkness between galaxies, invisible to our surveys but contributing significantly to the total stellar census.

Sirius has a companion star but not all stars have company, some form in complete isolation. (NASA, ESA and G. Bacon (STScI))

FACT 93

There are some black holes that are older than galaxies

In the primordial darkness of the early universe, before the first stars had ever begun to shine, black holes may have already been lurking in space. These ancient gravitational monsters, called *primordial black holes*, could be older than every galaxy, star and planet in existence and even witnessed the universe's violent birth and have been silently patrolling space for over 13 billion years.

Unlike the black holes we typically hear about, which form when massive stars collapse at the end of their lives, primordial black holes perhaps emerged from the extreme conditions that existed shortly after the Big Bang. In those first chaotic moments of cosmic history, when the universe was unimaginably hot and dense, tiny fluctuations in matter density could have triggered gravitational collapse on scales much smaller than stars.

These primordial black holes did not need the death of a massive star to form, they were born directly from the compressed matter of the early universe itself. Some might have been microscopic, smaller than atoms but containing the mass of asteroids. Others could have been as massive as mountains, packed into regions smaller than the full stop at the end of this sentence. Still others might have grown to stellar masses or beyond, becoming the seeds for the supermassive black holes we observe at the centres of galaxies today.

The existence of primordial black holes remains one of astronomy's most intriguing mysteries. We cannot see them directly because they do not emit light, and many would be far too small to detect with current technology. However, their gravitational effects might be all around us. Some scientists theorise

Simulated view of a black hole in front of the Large Magellanic Cloud, showing the event horizon.

that primordial black holes could partially explain dark matter – the mysterious invisible substance that makes up roughly 27% of the universe.

If primordial black holes exist in large numbers, they could have profoundly influenced cosmic evolution. During the universe's dark age, before the first stars formed, these ancient black holes might have served as gravitational seeds, attracting matter and helping to sculpt the cosmic web of structure we see today. They could have been the universe's first architects, laying the foundation for everything that followed.

Recent observations have provided tantalising hints that primordial black holes might be real. The gravitational wave detectors LIGO (the Laser Interferometer Gravitational-Wave Observatory) and Virgo have observed black hole mergers that seem unusually massive for stellar origin black holes, possibly suggesting a primordial origin. Additionally, some of the supermassive black holes observed in the early universe appear too large to have grown from stellar collapse alone, hinting that they might have started as primordial seeds.

The search for primordial black holes continues through multiple approaches. Astronomers look for their gravitational effects on light from distant stars, study the cosmic microwave background for evidence of their formation and analyse gravitational wave signals for signatures of primordial mergers.

FACT 94

Some exoplanets orbit dead stars

Inevitably, stars – and even our own Sun – will die, but something extraordinary is happening: planets are not only surviving their star's death but continuing to orbit the stellar corpses left behind. These resilient worlds, orbiting white dwarf stars and neutron stars, represent some of the most extreme planetary systems in the universe, defying our expectations about what can survive the death of a star.

Exoplanet HIP 65426b orbits a star called Najsakopajk but not all exoplanets orbit stars that are alive, some orbit dead stars. (ESO)

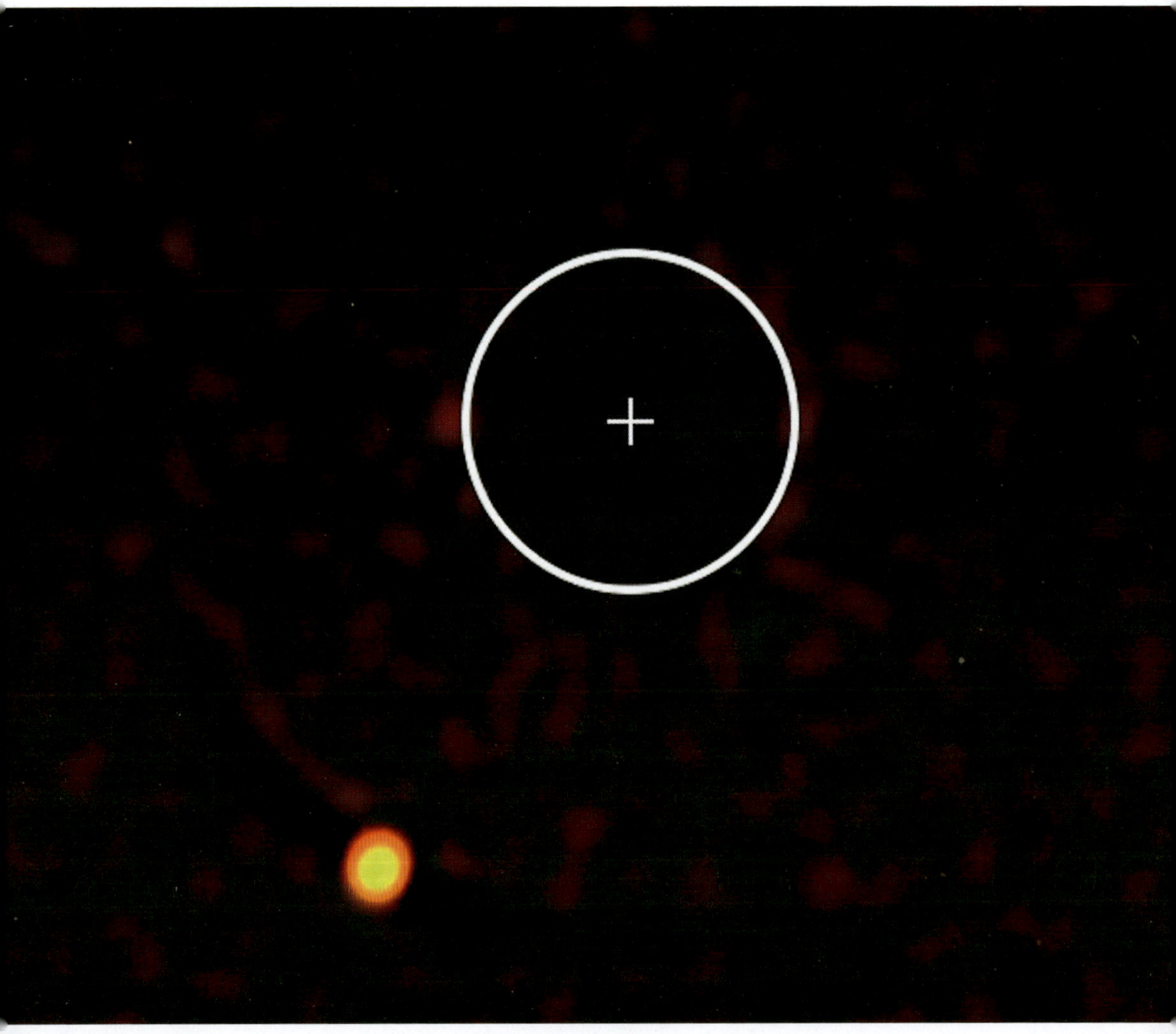

When a star dies, the process should theoretically destroy any nearby planets. As the most massive stars exhaust their nuclear fuel, they expand and explode as supernovae, potentially engulfing their inner planets in a fiery embrace. When smaller stars like our Sun reach the end of their lives, they shed their outer layers, creating stunning planetary nebulae, leaving behind only a hot, dense core called a white dwarf. These violent stellar death throes are generally expected to obliterate any orbiting worlds.

Yet against all odds, planets do seem to persist. Astronomers have discovered numerous exoplanets successfully orbiting these stellar remnants, creating some of the strangest planetary systems ever found. These worlds exist in environments where temperatures can reach thousands of degrees, radiation levels are extreme and the very fabric of space-time is warped by intense gravitational fields.

White dwarf planetary systems are particularly fascinating. These stellar corpses are incredibly dense: a teaspoon of white dwarf material would weigh as much as a large car on Earth. Despite their small size, roughly equivalent to Earth's diameter, they contain about half the mass of our entire Sun. Planets orbiting these compact objects experience gravitational forces unlike anything in our solar system.

Some of these hardy worlds may be survivors from the original planetary system, somehow managing to avoid destruction during their star's dramatic finale. They might have migrated outward as their star expanded, riding gravitational waves to safety before settling into new orbits around the stellar remnant. Others could be second-generation planets, forming from the debris left behind after the star's death.

The discovery of planets around neutron stars is even more remarkable. These ultra-dense stellar remains, formed when massive stars explode as supernovae, create some of the most extreme environments imaginable. A neutron star packs more mass than our Sun into a sphere just 20 kilometres across, creating gravitational fields so intense that escape velocity approaches a significant fraction of light speed.

The first confirmed exoplanets ever discovered were actually found orbiting a pulsar, a rapidly spinning neutron star that sweeps beams of radiation across space like a lighthouse. These pulsar planets experience radiation levels millions of times more intense than anything in our solar system, yet they continue to orbit their exotic host.

These extreme planetary systems challenge our understanding of planetary formation and survival. They suggest that planets are far more resilient than previously thought, capable of surviving some of the most extreme processes in the universe.

FACT 95

Organic molecules exist in interstellar gas clouds

The building blocks of life are floating freely throughout the Galaxy. Far from any planet or star, in the vast, seemingly empty spaces between star systems, astronomers have discovered complex organic molecules drifting through interstellar gas clouds. These cosmic chemistry labs are producing the same

LH 95 stellar nursery in the Large Magellanic Cloud (LMC). (NASA/ESA)

carbon-based compounds that form the foundation of life on Earth, suggesting that the ingredients for biology might be far more common in the universe than we ever imagined.

These molecular discoveries have revolutionised our understanding of space chemistry. What was once thought to be mostly empty vacuum punctuated by simple atoms has revealed itself to be a rich chemical soup containing hundreds of different organic compounds. Radio telescopes have detected everything from simple molecules like methanol and ammonia to complex structures containing dozens of atoms, including amino acids, the building blocks of proteins essential to all life.

The conditions in interstellar space seem hostile to chemistry at first glance. Temperatures hover just a few degrees above absolute zero, and the density is so low that molecules are separated by vast distances. Yet somehow, these frigid, sparse environments are conducting sophisticated chemical reactions. The secret lies in the unique physics of interstellar space. Unlike Earth-based chemistry, which relies on heat and collisions to drive reactions, space chemistry operates through different mechanisms. Cosmic rays, the high energy particles from distant stellar explosions, provide the energy needed to break and form chemical bonds. Dust grains act as tiny reaction hosts, allowing molecules to stick to their surfaces and interact in ways impossible in the gas phase alone.

Perhaps most remarkably, some of these interstellar clouds contain *prebiotic molecules*, complex organic compounds that are direct precursors to the molecules of life. Astronomers have detected glycine, the simplest amino acid, as well as various sugars and even the components of nucleotides that form DNA and RNA. These discoveries suggest that the chemical pathway from simple atoms to life's complex molecules may begin long before planets even form.

The implications for astrobiology are staggering. If organic molecules are being manufactured throughout the Galaxy in interstellar space, then every new solar system may start with a ready-made supply of life's building blocks. When planets form from these chemically enriched gas clouds, they inherit molecules that could jump-start the development of life.

This chemical process also explains how organic compounds reach places where life might take hold. Comets and asteroids, formed from the same interstellar material, become delivery vehicles for these prebiotic molecules. When they impact planetary surfaces, they seed worlds with the chemical ingredients necessary for life to begin.

The discovery that space is teeming with organic molecules fundamentally changes how we think about life's origins. Rather than life emerging from purely terrestrial chemistry, it appears that the universe has been preparing the ingredients for biology for billions of years, spreading them throughout the Galaxy.

FACT 96

The universe may be shaped like a doughnut

The universe might be shaped like a doughnut, wrapped around itself in a mind-bending topology that would make travelling in a straight line eventually bring you back to where you started. This bizarre possibility, known as a *toroidal universe*, represents one of the most intriguing theories about the overall shape and structure of our cosmos, challenging our intuitive understanding of space itself.

Most people imagine the universe as an infinite expanse stretching endlessly in all directions that keeps expanding forever. However, mathematicians and cosmologists have discovered that the universe could have a finite size while still appearing infinite to anyone living inside it. Just as the surface of Earth is

The curvature of space is now thought to be represented by the second image in the diagram and shaped like a donut

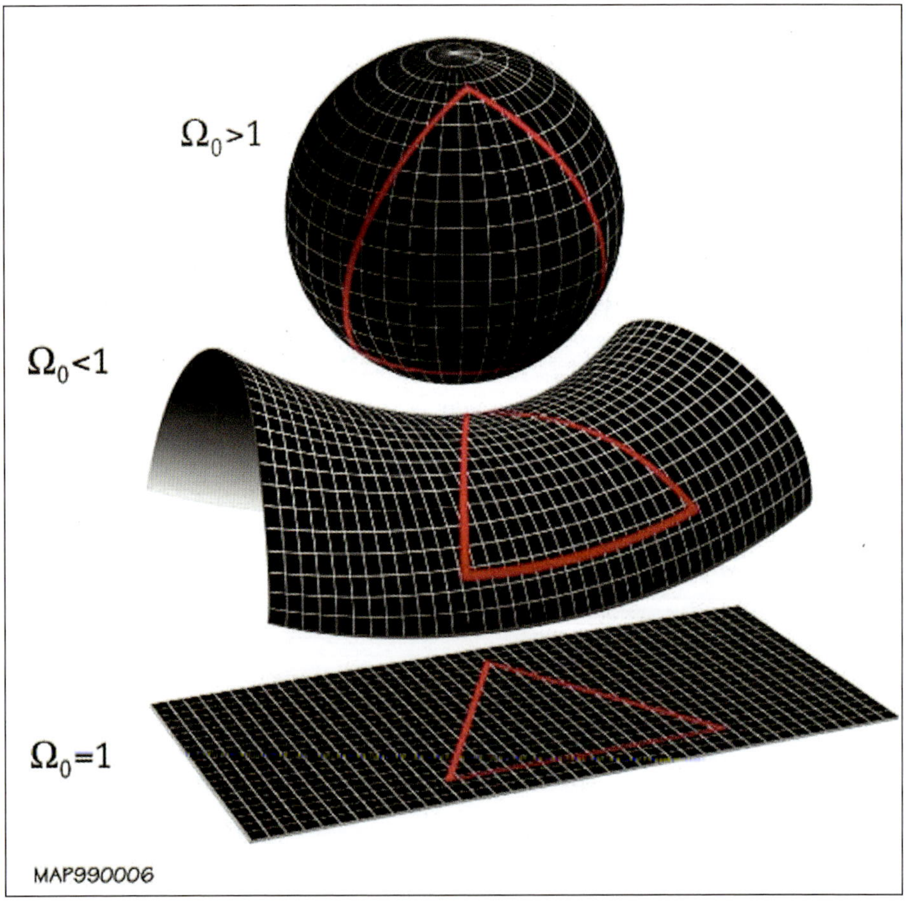

finite but has no edges, you can, after all, travel around the globe and end up back where you started. It seems the universe might curve back on itself too, but in higher dimensions.

The doughnut-shaped universe, technically called a *3-torus*, would create the ultimate cosmic illusion. If you could somehow travel far enough in any direction, you would eventually return to your starting point from the opposite direction. Imagine looking through a powerful telescope and seeing the back of your own head in the distance – that's the strange geometry a toroidal universe would create.

This topology would produce fascinating observational effects. Astronomers might see multiple images of the same distant galaxy scattered across the sky, like looking at yourself in a hall of mirrors. The cosmic microwave background radiation, the afterglow of the Big Bang, might actually show repeating patterns that could serve as evidence for this doughnut-like structure.

Several lines of evidence have led scientists to consider this possibility seriously. Observations of the cosmic microwave background suggest that space might be slightly curved, and some measurements hint at anomalies that could be explained by a finite, wrapped universe. Additionally, computer simulations show that a toroidal universe could solve certain puzzles about why the universe appears so uniform in all directions, as observations support.

The doughnut universe theory also addresses a fundamental question: what lies beyond the edge of space? If the universe has finite size but is shaped like a doughnut, there is no edge, space and time simply curves back on itself. This eliminates the conceptual problem of what exists 'outside' the universe, since there would be no outside.

Detecting whether we live in a great big doughnut poses significant challenges. The evidence would be subtle, requiring careful analysis of patterns in the cosmic microwave background or the distribution of galaxies across vast distances. Some researchers have claimed to find hints of such patterns, though the results do still require confirmation.

If confirmed, a doughnut-shaped universe would revolutionise our understanding of cosmic geography. It would mean that the universe, while vast beyond human comprehension, is ultimately finite and self-contained. Every point in space would be connected to every other point through the underlying topology, making the cosmos both intimately connected and elegantly self-contained.

FACT 97

There is an exoplanet where it rains molten glass

In the exoplanet catalogue of alien worlds, few planets embody nightmare more completely than HD 189733b. This distant exoplanet, located 63 light years from Earth, experiences weather so extreme it defies earthly comprehension: sideways rain made of molten glass, driven by supersonic winds screaming across the planet's surface at 7,000 kilometres per hour, nearly seven times faster than the speed of sound.

HD 189733b belongs to a class of planets known as 'hot Jupiters', gas giants that orbit perilously close to their host stars. While Jupiter in our solar system takes 12 Earth years to complete one orbit, HD 189733b races around its star in just 2.2 Earth days. This unimaginable orbital speed keeps the planet locked in a gravitational embrace where one side perpetually faces its star while the other remains in eternal darkness.

The planet's proximity to its star creates a hellish greenhouse effect. Surface temperatures soar to over 1,000°C, hot enough to melt copper and approaching the melting point of silver. But temperature is just the beginning of this world's meteorological madness. The extreme heating creates colossal atmospheric circulation patterns that generate the most violent weather system ever discovered.

The infamous glass rain forms through a process both fascinating and terrifying. The planet's atmosphere contains silicon compounds that, under the intense heat and pressure, condense into tiny particles of silicate, essentially microscopic glass beads. These glass droplets form clouds that would make any earthly storm seem like a gentle spring shower by comparison.

What makes this weather truly alien is its behaviour. On Earth, rain falls downward due to gravity. On HD 189733b, the phenomenal wind speeds blast these molten glass particles horizontally across the planet's surface. The extreme temperature difference between the planet's day and night sides drives these incredible wind speeds. The star-facing hemisphere bakes in perpetual heat while the dark side freezes in permanent midnight. This creates massive pressure differentials that generate atmospheric circulation patterns capable of moving at velocities that would make Earth's most powerful hurricanes seem like gentle breezes.

Astronomers discovered these extreme conditions using sophisticated spectroscopy techniques, analysing how starlight filters through the planet's atmosphere as it passes in front of its host star. The distinctive signatures of silicon compounds and the Doppler shift of atmospheric gases revealed both the composition and motion of HD 189733b's deadly weather systems.

Simulation of a 'hot Jupiter' planet known as HD 80606b. (NASA/JPL Caltech)

FACT 98

Stars can be 'spaghettified' by black holes

In one of the most violent and bizarre phenomena in the universe, black holes can literally stretch entire stars into long, thin strands resembling spaghetti before tearing them apart completely. This process, appropriately named 'spaghettification', represents one of the most extreme demonstrations of gravity's immense power, capable of reducing massive stellar objects to streaming ribbons of superheated gas.

The process begins when a star ventures too close to a supermassive black hole. These gravitational monsters, millions or billions of times more massive than our Sun, lurk at the centres of most galaxies, waiting like traps for unwary stars to wander within their deadly reach. Once a star crosses a critical boundary called the *tidal disruption radius*, its fate is sealed.

Spaghettification occurs because gravity follows an inverse square law: its strength decreases rapidly with distance. When a star approaches a black hole,

Spaghettification of a star by a supermassive black hole. (Daniel James Price)

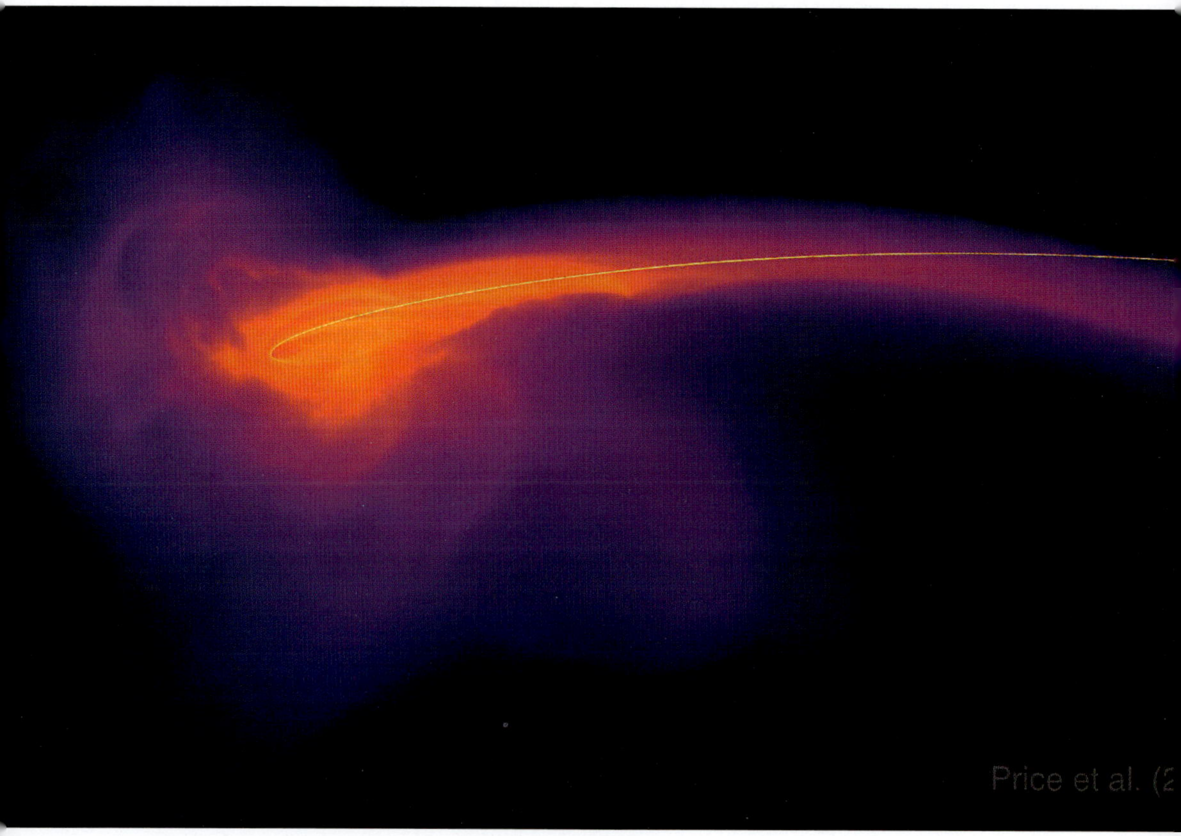

the gravitational force pulling on the star's near side becomes dramatically stronger than the force acting on its far side. This difference in gravitational pull, called a tidal force, begins to stretch the star along the direction pointing towards the black hole while simultaneously compressing it from the sides.

As the star stretches, its material heats up due to compression and friction, glowing brilliantly as it spirals toward its doom. This creates a spectacular light show visible across interstellar distances. The process typically takes months or years to complete, during which half the star's material falls into the black hole while the other half gets ejected back into space at tremendous speeds.

The discovery of tidal disruption events has changed our understanding of both black holes and stellar dynamics. It is thought that these events occur roughly once every 10,000 to 100,000 years in any given galaxy, making them relatively rare but incredibly valuable for studying black hole behaviour. When astronomers observe the characteristic flare of a spaghettified star, they can measure the black hole's mass, spin and other properties.

Recent observations have captured these events in unprecedented detail. The European Southern Observatory (ESO) and other facilities have witnessed stars being stretched and torn apart in real time, providing insights into how matter behaves under the most extreme gravitational conditions imaginable. Some disrupted stars create bright flares that outshine entire galaxies for months.

Spaghettification is not limited to stars – any object that ventures too close to a black hole faces the same fate. Planets, asteroids, entire solar systems and even an unwary astronaut could be stretched into space pasta if they wandered into a black hole's gravitational web.

FACT 99

Some white dwarfs can reignite and explode

The stellar graveyard is not as peaceful as it might seem. White dwarfs, the dense, cooling cores of dead stars that should theoretically fade quietly into the darkness, can sometimes experience violent resurrections, reigniting in spectacular explosions that briefly outshine entire galaxies. These stellar zombies demonstrate that death is not always final, even in the realm of the stars.

White dwarfs are among the most common stellar corpses in the universe. When stars like our Sun exhaust their nuclear fuel, they shed their outer layers and leave behind a hot, dense core roughly the size of Earth but containing

A comparison between the white dwarf IK Pegasi B (centre), its A-class companion IK Pegasi A (left) and the Sun (right). This white dwarf has a surface temperature of 35,500 K. (R.J. Hall)

about half the Sun's mass. These objects are incredibly dense and slowly, over billions of years, will cool and fade, radiating their heat into space until they become cold, dark objects called black dwarfs. However, the universe has other plans for some of these stellar remnants. When a white dwarf has a companion star in a binary system, it can begin a process that leads to one of the most powerful explosions in the cosmos.

The key to this stellar resurrection lies in mass transfer. If the companion star ventures too close or expands as it ages, the white dwarf's intense gravitational field can strip material from its neighbour. This stolen matter, primarily hydrogen and helium, accumulates on the white dwarf's surface like snow building up on a rooftop. But unlike ordinary snow, this stellar material is under extreme pressure and temperature.

As more material accumulates, the bottom layers get compressed and heated to incredible temperatures. Eventually, the pressure and heat reach a critical threshold where nuclear fusion suddenly reignites on the white dwarf's surface. This is not the gradual, controlled fusion that powered the original star, but an explosive thermonuclear runaway reaction that detonates like a hydrogen bomb.

This explosion, called a nova, can increase the white dwarf's brightness by factors of 100,000 to millions, making it visible from vast distances across the galaxy. The blast hurls the accumulated material back into space at speeds of thousands of kilometres per second, creating brilliant shells of glowing gas that expand outward.

Some white dwarfs experience even more dramatic fates. If they accumulate enough material to approach the *Chandrasekhar limit*, roughly 1.4 times the mass of our Sun, the entire white dwarf can explode as a 'Type Ia' ('type one-A') supernova. These explosions are so powerful they completely destroy the white dwarf, leaving nothing behind but an expanding cloud of stellar debris enriched with heavy elements forged in the explosion.

These explosions play crucial roles in the chemistry and evolution of the cosmos. Novae and supernovae scatter heavy elements throughout the Galaxy, seeding space with the carbon, oxygen, silicon and iron necessary for rocky planets and life itself. The process is an exact one too, which means they shine in a way that enables us to use them as an aid to measure distances in the universe.

FACT 100

There are supersonic winds on exoplanets

Far beyond our solar system, alien worlds experience weather patterns that would make Earth's most powerful hurricanes seem like gentle breezes. Scientists have discovered that many exoplanets that orbit other stars are battered by winds so fierce they break the sound barrier, creating supersonic storms that reshape entire atmospheres.

The most extreme examples of these cosmic tempests occur on a class of planets called 'hot Jupiters'. These massive gas giants orbit incredibly close to their host stars, sometimes completing a full orbit in just a few days. The intense stellar radiation heats these worlds to scorching temperatures, creating the perfect conditions for atmospheric chaos.

One of the best studied examples is HD 209458b, a planet roughly the size of Jupiter but orbiting so close to its star that surface temperatures reach over 1,000°C. Here, winds have been measured exceeding 7,000 kilometres

The exoplanet HD 209458 b seen here in comparison to Jupiter, experiences supersonic wind speeds. (Aldaron)

per hour, speeds that dwarf anything found in our solar system. To put this in perspective, these winds are more than twice as fast as the speed of sound and roughly 20 times stronger than the most powerful hurricanes on Earth.

What makes these supersonic winds possible is the extreme temperature difference between a hot Jupiter's day and night sides. Since these planets are tidally locked, meaning the same side always faces their star, one hemisphere experiences eternal day while the other remains in perpetual darkness. The day side can be thousands of degrees hotter than the night side, creating a massive pressure gradient that drives the atmosphere into violent motion.

These extreme winds redistribute heat around the planet's atmosphere, in the same way as winds do here on Earth, but to far more extreme levels. Hot gas from the day side gets swept around to the night side, while cooler material flows back to replace it. This process prevents the night side from becoming as cold as it would otherwise be, moderating the planet's overall temperature structure.

The discovery of supersonic winds on exoplanets has changed our understanding of atmospheric dynamics beyond Earth. Using advanced telescopes and spectroscopic techniques, astronomers can detect these winds by observing how starlight filters through a planet's atmosphere as it passes in front of its host star.

These findings have important implications for the search for habitable planets. Understanding how extreme atmospheric conditions develop helps researchers identify which types of worlds might be suitable for life. While 'hot Jupiters' themselves are far too hostile for any known form of life, studying their atmospheres provides valuable insights into the wide range of planetary climate systems.

FACT 101

The Milky Way has a 'fossil' galaxy inside it

Hidden within the familiar spiral arms of our Milky Way Galaxy lies the ancient remains of a collision that occurred billions of years ago. Like palaeontologists discovering dinosaur bones buried in rock layers, astronomers have uncovered evidence of a small galaxy that was devoured by our own, leaving behind a trail of stellar fossils that continue to orbit within our galactic home.

This remarkable discovery began with data from the European Space Agency's (ESA) *Gaia* spacecraft, which has been precisely measuring the positions and movements of over a billion stars in our Galaxy. By analysing these stellar motions, astronomers noticed something unusual – a large group of stars moving in distinctly different patterns from the rest of the Milky Way's population.

It is now believed that a small galaxy, named Gaia-Enceladus, collided and merged with the Milky Way billions of years ago and it is these stars that are now being observed. Named after both the *Gaia* mission that detected it and Enceladus, a giant from Greek mythology, this ancient galaxy was roughly one fourth the mass of the Milky Way when the collision occurred approximately 10 billion years ago.

The evidence for this merger is written in the stars themselves. Its stellar remnants still orbit within our Galaxy, following elongated, tilted paths that are completely different from the circular orbits of stars born within the Milky Way. These stars act like a galactic fossil record, preserving the memory of their original galaxy's motion billions of years after their home was destroyed.

What makes Gaia-Enceladus particularly fascinating is how it transformed our Galaxy's structure. Before this collision, the Milky Way was likely a smaller, thinner disc galaxy. The violent merger not only added mass and stars but also disrupted the galaxy's original architecture. The gravitational chaos scattered existing stars into new orbits and helped create the Milky Way's thick disc and stellar halo, the sparse collection of ancient stars that surrounds our Galaxy's main body.

It is possible to identify these immigrant stars by studying their chemical composition. Stars from Gaia-Enceladus contain different ratios of heavy elements compared to native Milky Way stars, reflecting the unique star formation history of their original galaxy. This chemical fingerprinting allows us to distinguish between stars that formed within our Galaxy and those that arrived as a galactic merger.

The discovery also provides insights into the Milky Way's future. Astronomers predict that our Galaxy will eventually collide with the nearby Andromeda Galaxy in approximately 4.5 billion years. By studying the remnants of Gaia-Enceladus, we can better understand what to expect from this future galactic encounter.

Artist's impression of debris of the Gaia-Enceladus galaxy. (ESA)

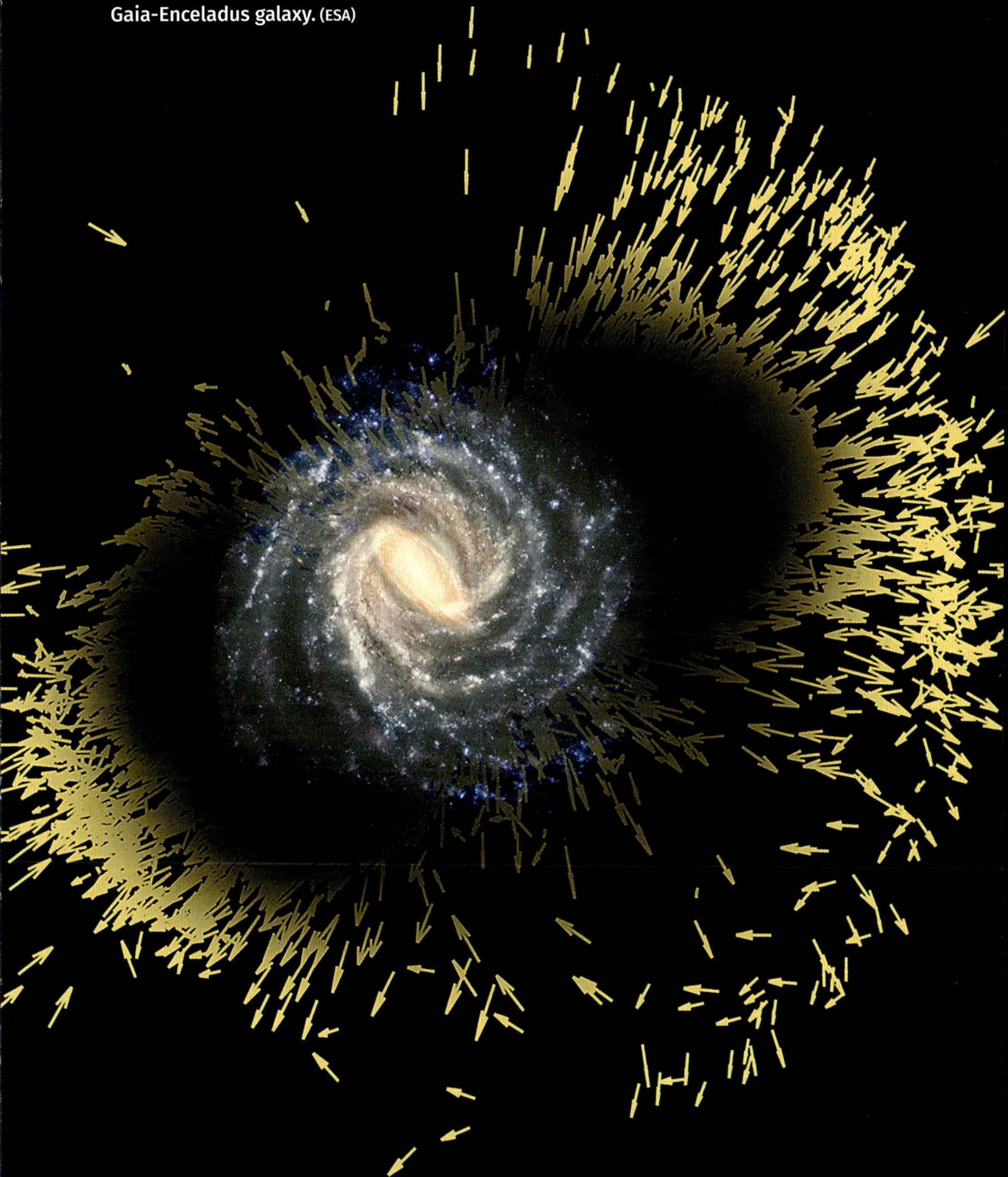